A Manual of the
SLIDE RULE
Its History, Principle and Operation

BY

J. E. THOMPSON, B.S. *in* E.E., A.M.
Department of Mathematics, Pratt Institute, Brooklyn

TWELFTH PRINTING

WILDSIDE PRESS

NEW YORK
D. Van Nostrand Company, Inc., 250 Fourth Avenue, New York 3

TORONTO
D. Van Nostrand Company (Canada), Ltd., 228 Bloor Street, Toronto

LONDON
Macmillan & Company, Ltd., St. Martin's Street, London, W.C. 2

Copyright, 1930
By D. VAN NOSTRAND COMPANY, Inc.

All rights reserved.

This book, or any parts thereof, may
not be reproduced in any form without
written permission from the publisher.

First published, Sept. 1930

Reprinted, March 1931, October 1937
November 1941, June 1942
February 1943, October 1943, September 1944
July 1945, March 1947, January 1949
April 1951

PREFACE

Accounts of the slide rule are usually confined to encyclopedia articles, supplementary chapters of books on other subjects, and manufacturers' booklets of directions for its use. The first two are usually sketchy and incomplete and intended for general information rather than for specific instruction. The manufacturers' booklets are well suited to their purpose, which is to give explicit directions for the mechanical operation of some one form of the rule and examples illustrating its use in particular calculations, but are not intended as general books on the slide rule.

The present manual is intended to supply in a uniform presentation an account of the history, principles and practical uses of the slide rule. Among its specific aims are: to show the unity of the principle of operation of all slide rules, to explain this principle in a manner which shall be easily understood, to give in clear and simple language explicit and inclusive instructions for the use of all the standard forms, and to supply a fairly complete account of the history of the slide rule and a description of the more usual of the special forms of the rule. The material in Chapter IV is intended to be more suggestive than exhaustive and that in Chapter V is intended more as supplementary general information than as instruction in the use of those rules described.

It is hoped that Chapters II, III and IV will be of value to all users of the slide rule and suitable for self instruction, and that Chapters I and V will be of interest to all who pay any attention to the slide rule in any way. In the preparation of Chapters II, III and IV, great care has been used to make the descriptions and explanations complete, simple and clear. The style adopted has been found very useful and effective in the author's teaching experience with users of the slide rule, among whom have been numbered high school and college students, practical industrial workers and clerical workers. It is for these classes of readers that the book is designed.

It is a pleasure to acknowledge a great and peculiar obligation to Mr. W. H. H. Cowles of Pratt Institute. He originally undertook the preparation of the book and had prepared a tentative list of chapter headings and outlined the scope of the book when new duties attendant upon a change in his relation to the Institute necessitated his relinquishing the work. The present writer undertook the work at Mr. Cowles' suggestion and has since profited by frequent discussions with him.

The kindness and courtesy of Keuffel & Esser Company has made possible the inclusion of almost all the material of Chapters IV and V, and the plates for a number of the illustrations were graciously loaned by them. To this company the obligations of the author and publisher are hereby acknowledged with thanks.

As the proof has been read by the author alone errors will undoubtedly be found. Notices of these will be gladly received by him or the publisher.

BROOKLYN, N. Y. J. E. THOMPSON.
September, 1930

CONTENTS

CHAPTER I

HISTORY OF THE SLIDE RULE

ARTICLE		PAGE
1.	Introduction.—Logarithms	1
2.	Logarithmic Multiplication and Division and the Invention of the Slide Rule	3
3.	Improvement and Development of the Slide Rule in England	6
4.	The Slide Rule in France	9
5.	The Mannheim Slide Rule	13
6.	Recent Developments	15
7.	The Slide Rule in the United States	16

CHAPTER II

THEORY AND OPERATION OF THE MANNHEIM SLIDE RULE

A. THEORY OF THE RULE

8.	Some Properties of Exponents	21
9.	Logarithms	23
10.	Principle of the Slide Rule	25
11.	Application of the Slide Rule Principle	30
12.	The Standard Mannheim Slide Rule	33

B. OPERATION OF THE RULE

13.	Division and Reading of the Scales	35
14.	Multiplication with the Slide Rule	40
15.	Division and Reciprocals	48
16.	The Decimal Point in Slide Rule Multiplication and Division	53

ARTICLE	PAGE
17. Combined Multiplication and Division	58
18. Proportion	62
19a. Squares, Cubes and Roots	63
19b. Use of Gauge Points on the A and B Scales	73
20. The Inverted Slide	74
21. The Scale of Equal Parts (L) and Its Use	78
22. The Sine Scale (S) and Its Use	80
23. The Tangent Scale (T) and Its Use	84

CHAPTER III
MODIFIED FORMS OF THE MANNHEIM RULE AND THEIR USE

A. THE POLYPHASE MANNHEIM SLIDE RULE

24. The Polyphase Mannheim Slide Rule	88
25. The Inverted C Scale and Its Use	90
26. The Cube Scale and Its Use	95

B. THE POLYPHASE DUPLEX SLIDE RULE

27. The Duplex Slide Rule and the Folded Scales	99
28. Operation of the Polyphase Duplex Slide Rule	102

C. THE LOG-LOG DUPLEX SLIDE RULE

29. Natural Logarithms and Exponentials	109
30. Exponentials, Powers and Roots and the Slide Rule	115
31. The Log-log Scale and the Log-log Duplex Slide Rule	117
32. Use of the Log-log Scales	121

D. THE SLIDE RULE IN TRIGONOMETRY

33. Introductory	133
34. Radian and Degree Angle Measure Conversion	134
35. Solution of Right Triangles	135
36. Solution of Oblique Triangles	140
37. Slide Rule Check of Logarithmic Solutions	146
38. Interpolation of Logarithms	147

CHAPTER IV
TYPICAL PROBLEMS AND SLIDE RULE SETTINGS

		PAGE
A.	Introduction	151
B.	Typical Problems and Settings for A, B, C, and D Scales	152
C.	Tables of Conversions and Gauge Points for C and D Scales	164
D.	Settings for Polyphase Mannheim Rule .	171
E.	Settings for Polyphase Duplex and Log-log Duplex Rules	177
F.	Settings and Typical Problems Involving the Log-log Scales	185

CHAPTER V
SPECIAL FORMS OF THE SLIDE RULE

ARTICLE
39. Introduction 204
40. Slide Rules of Mannheim Form with Special Scales . 206
41. Duplex Slide Rules with Special Scales 210
42. Circular Slide Rules 214
43. Cylindrical Slide Rules 217

CHAPTER I

HISTORY OF THE SLIDE RULE

1. Introduction — Logarithms. — The modern slide rule is frequently thought of as a modern invention but in its earliest forms it is several hundred years old. As a matter of fact, it can hardly be called an invention in the ordinary sense, but is rather an outgrowth of certain ideas in mathematics. In this chapter we shall trace briefly the origin of the slide rule in these ideas and the development of its various forms into the familiar instrument of today.

Insofar as the slide rule can be called a distinct invention it originated as a scale on which were laid off lengths representing the logarithms of numbers so that multiplication and division could be performed mechanically by finding the sums and differences of lengths on this scale. But such a scale is simply a condensed table of logarithms, so the history of the slide rule in reality begins with the invention of logarithms.

Logarithms were invented by John Napier, Baron of Merchiston in Scotland, and his discovery was first publicly announced in 1614. In this announcement Napier stated his purpose in the following words: "Seeing there is nothing (right well beloved Students of Mathematics) that is so troublesome to mathematical practice, nor doth more molest and hinder calculators, than the multiplications, divisions, square and cubical extractions of great numbers, which beside the tedious expense of time are for the most part subject to slippery errors, I began therefore to consider in my mind by what certain and ready art I might remove those hindrances." The slide rule represents the most "certain and ready" fulfillment of this purpose, and is the final result of his labors. Although the completed invention of logarithms was not publicly announced until 1614, Napier had privately communicated a summary of his results to the Danish astronomer, Tycho Brahe, in 1594. Thus in its fundamental principle the slide rule dates back nearly three and one half centuries, beginning only a hundred years after the discovery of America.

In his book of 1614 on logarithms Napier explains their nature by a comparison between corresponding terms of an arithmetical and a geometrical progression. He gives tables of the logarithms of the sines and tangents of angles calculated to seven decimal places. His definition of the logarithm of a number N was what we should express, according to our present definition, by the

quantity $10^7 \log_e \left(\frac{10^7}{N}\right)$, where $e = 2.71828$ approximately.
The transformation to the form described in the present day definition was made by Henry Briggs, Professor of Mathematics at Oxford University, who in 1617 brought out the first table of ordinary logarithms of the natural numbers from 1 to 1000 to the base 10 calculated to fourteen decimal places. These were later extended and now tables of all the numbers from 1 to 108000 are common and some have been computed to sixty decimal places.

2. **Logarithmic Multiplication and Division and the Invention of the Slide Rule.** — In order to carry out the multiplication of numbers we add their logarithms and take the "antilog" of the sum; thus if $x = AB$ then $\log x = (\log A + \log B)$ and the product $x =$ antilog $(\log A + \log B)$, to any base whatever. Similarly if $y = \frac{A}{B}$, $y =$ antilog $(\log A - \log B)$. The logarithms of A and B are found in a table of logarithms and the antilogarithm of the sum or difference is found in the same table. This process may also easily be carried out mechanically in the following manner: Lay off to a convenient scale on a straight line the logarithms of the natural numbers and instead of marking log 1, log 2, log 3, etc., on the scale simply mark 1, 2, 3, etc. Then to find $x = AB$ measure by a pair of dividers the scale length of log B,

that is, the length from the beginning of the scale to the point marked B. Now lay off on the scale with the dividers this same length beginning at the point marked A. The total length is equal to ($\log A + \log B$) and the number at the end of this total length on the scale is x. Such a logarithm scale, laid off on a strip of wood, was first constructed by Edmund Gunter, Professor of Astronomy at Gresham College, in London, in 1620. This device was known as Gunter's Scale or Gunter's Rule. Obviously Gunter's Rule may be used for division by measuring the difference between the scale lengths for dividend and divisor, that is, $\log A - \log B$.

The process just described can be much more easily and quickly carried out if we use two similar scales and instead of measuring and adding or subtracting the lengths on one scale, simply lay one upon or beside the other, with the beginning of the second scale placed at the point corresponding to A on the first. Then the point corresponding to B on the second will be beside the point on the first which is at the end of a length equal to the sum of the two lengths, that is, the point corresponding to $x = AB$. Similarly the length equal to $\log A - \log B$ may be found and the point at the end of this length will be $y = \dfrac{A}{B}$. Such a combination of Gunter's Rules was first made by Rev. William Oughtred who lived near London, and was first made public by William Forster, a teacher of mathematics in London, in 1632, when he stated that he first saw the double rule at Oughtred's home in 1630 at which time Oughtred told him he had constructed the double rule several years previously. Oughtred's double rule was made in the straight form with one rule to be held against and *slide* along the other by hand, and also in circular form on cardboard, one scale being on the perimeter of a circular disc which was pivoted at the center of a similar and larger disc and turned upon it, so that the smaller circular scale lay just inside of and in alignment with the larger, which was laid off on the perimeter of the larger disc. These rules carried scales of the logarithms of the natural and trigonometric numbers. William Oughtred is thus the inventor of the *slide rule* in both the straight and circular forms.

The rules were for the first time made up of three strips of wood with the two outer strips held together by bridging cleats and the third strip sliding freely between them edge to edge, by Seth Partridge, a surveyor and teacher of mathematics, in 1657. This combination rule carried the number and trigonometric scales on both faces and was thus the first complete slide rule. It was as a matter of fact a complete *duplex slide rule*, which is thus now about 270 years old.

3. Improvement and Development of the Slide Rule in England. — The name "sliding-rule" (later abbreviated to "slide rule") was first used by Thomas Everard, a mechanic and excise officer at Southampton, who in 1683 brought out rules adapted for the computations involved in determining cubic contents of measuring vessels and containers of market commodities. On his rules were marked so-called "gauge points" or numbers often used in slide rule settings involving the standard measurements and measuring unit conversion factors. These were the first gaugers' or commercial slide rules. On these rules appeared for the first time scales which could be used for finding the square and cube roots of numbers. The first slide rule scales designed especially for the direct reading of the squares and cubes and square and cube roots of numbers were made by one John Warner in the year 1722.

In the year 1697 William Hunt, of whom very little is known, brought out a slide rule having in addition to the usual scales a scale giving the areas of circles of known diameters, one for finding the perimeters of ellipses of known axes, and a scale for finding the length or width of a rectangle of unit area when the width or length respectively was known. Since when the area is unity, 1, $WL = 1$ and hence $L = \frac{1}{W}$ or $W = \frac{1}{L}$, this scale was in reality a scale of *reciprocals*, now known as the "inverted scale." The reciprocal or inverted scale was first placed on the rule as a simple regular scale running in the reverse direction by William Hyde Wollaston soon after the year 1797. Incidentally, this same Wollaston brought out in 1815 a slide rule with scales specially adapted for the calculations involved in chemistry.

A development or use of the slide rule which is not commonly known but which later led to a very important improvement was made by Sir Isaac Newton, the inventor of the infinitesimal calculus, discoverer of the law of gravitation, and founder of modern mechanics and dynamical astronomy. In his work in the theory of algebraic equations he devised a method of solving cubic (third degree) equations in which three movable slide rule scales were laid side by side and a separate straight edge laid across them to bring together or in line three certain numbers on scales not contiguous to one another. This is the first recorded use of what is now known as a *runner*. In discussing Newton's work this device was made more definite and described as a "marking line moving parallel to itself" by E. Stone in 1726. This moving marking line was first definitely attached to the slide rule in a regular manner as a mechanical part of the rule itself in 1775 by John Robertson, teacher of mathematics at the Royal Academy at Portsmouth and later Librarian of the Royal Society of London.

The circular slide rule invented in England by Oughtred about or before 1630, as stated above, was independently re-invented in France in 1716 by Jean Baptiste Clairaut, the father of the famous mathematician of that name, but was used hardly at all until 1748. In that year it was brought out in a more complete and convenient form by George Adams, manufacturer of mechanical instruments for King George III. Circular and spiral forms of the scales were much improved toward the end of the 18th century by William Nicholson (see below).

Before 1775 the various slide rules in use in England and the American colonies were very poorly constructed and ruled. In the year 1707 one John Ward in a work on practical mathematics and gauging wrote that results obtained by their use were so inaccurate that the pen should always be used in such work, though he later stated that he did not refer to the long rules. About 1775 James Watt, one of the inventors of the steam engine, had made to his own order and with great care excellent rules for his own use in engine design. This seems to have had the double effect of setting a high standard of workmanship and populaiizing the use of slide rules among English engineers.

About or soon after this time William Nicholson (born 1753, died 1815) publisher and editor of "Nicholson's Journal," a kind of technical journal, began to devote most of his time to the study and improvement of the slide rule and used his "Journal" to advertise various rules and promote their use. He especially improved the circular and spiral forms of rules. In "Nicholson's Journal" there was announced in 1817 a new development which, while not used for a long time, was an important development; it was due to Sylvanus Bevan. This was the introduction of what is now known as the *folded scale*. This consisted of a scale of the same length and graduations as the usual logarithmic scale but divided into two parts so that it began at the left-hand end of the rule at a point a little past 3 and proceeded to 10 near the center, and beginning again at that point with 1 proceeded to the same number at the right-hand end as was placed at the left. Thus the normal scale was "folded" at the middle so that the ends were together; it was then parted at the folding point and these separated section ends were used as the ends of the new scale.

After Nicholson's day the slide rule was neglected and fell into disuse in England. France became the leading country in its development and production.

4. The Slide Rule in France. — Gunter's scale was introduced into France in 1624 by Edmund Wingate, an early and prolific English writer on the slide rule, who was for a long time believed by many to have been the inventor of the "sliding-rule." It has been definitely shown, however, that he merely added a scale of equal parts to Gunter's Rule so as to show the logarithms of the numbers on Gunter's scale, and that he used the slide rule only after it was made public by Oughtred.

For a long time the French did little in the way of development or improvement and merely used the English rules first advertised and taught by Wingate; even these were not much used or known before about 1700. In fact, this was the case all over the continent of Europe, and the first mention of the slide rule by a German writer is in a Latin work on mathematical instruments written by one Biler published in 1696. Throughout Europe, slide rules were not popular during the 17th and 18th centuries but, as we have already seen, they were popular in England, where decimal fractions and logarithms were much more a part of elementary mathematical education than on the continent.

The time of the French Revolution and the Napoleonic Wars, however, was a period of intense intellectual activity in France. Some of the world's greatest mathematicians and mathematical physicists (Lagrange, Laplace, Legendre, John Bernoulli the younger, Fourier, Poisson, Ampère, etc.) flourished there in that period. Besides the great development in mathematical analysis there were introduced a new calendar and a new system of angle measurement, though these were never widely used, the decimal (so-called "metric") system of measurement was invented and established, new and wonderful mathematical tables and methods of computation were produced, and the slide rule came into use and was improved in design and brought to a high state of mechanical perfection. Toward the middle of the century just dawning France became the center of activity in the study and production of the slide rule and the government made knowledge of the slide rule a requirement for admission into all technical public service.

Other European countries still lagged behind, however, although in 1843 an Austrian, L. G. Schulze von Strasnitzki, Professor at the Royal Institute at Vienna, made and used a slide rule eight feet long for demonstration and instruction purposes. Such a rule was at one time soon afterward used by the firm of Tavernier-Gravet in Paris and is now supplied in the United States by Keuffel & Esser.

In 1716, as we have seen, the circular rule was independently re-invented in France and circular rules have been popular there for a long time, but beyond the general mechanical improvement and development the greatest distinct contribution to the slide rule was the *log-log* scale, which was invented in 1815 by Peter M. Roget, a French physician, and reported by him in that year to the Royal

Society of London. Roget called his scale a "logometric" scale and pointed out that it could be used for raising any number to not only the second or third but to any power (within any desired range determined by the length of the scale) integral or fractional, and for taking corresponding roots. He called attention to its use in many calculations involving exponentials (population increase, compound interest, probability, theory of logarithms, etc.) but such computations were not then of much practical use and the log-log scale fell into disuse and was forgotten. It was later independently re-invented in France by Burdon, who proposed arrangements for finding x and y in such simultaneous combinations as $xy^m = a$, $xy^n = b$. The case of $x = \sqrt[n]{a}$ when a is a fraction was first provided for by F. Blanc in Germany before 1900. Due to the requirements of calculations in thermodynamics, hydraulics, electricity, etc., the various forms of log-log scales were independently re-invented in England by several, including Professor John Perry, the famous engineering mathematician and teacher. One of them, Professor C. S. Jackson, says, "The use of a log-log scale for numbers less than unity was one fondly thought new, but in this idea we were all anticipated by Blanc," and of course in the original idea of the log-log scale Blanc, as well as Burdon and the others, was long anticipated by Roget.

Log-log scales designed for the solution of algebraic equations up to and including the quintic, carrying out Newton's idea, were introduced by Furle in Germany in 1899 and scales for use with complex quantities $a + b\sqrt{-1}$ were designed by R. Mehmke in Germany soon after 1900.

The next distinct advance after the invention of the log-log scale in 1815 was made by Lenoir in France in 1821. He constructed after designs of Jamard and Collardean a machine for dividing the scales, which work had previously been done by hand. He developed this dividing engine so that it would mark the scale divisions simultaneously upon eight rules 25 centimeters long, executing very fine work.

The most important step in the evolution of the slide rule after Roget's invention was also made in France, in 1850, and is due to Lieutenant Amédée Mannheim. This development has since proved to be of such importance that it will be considered separately in the next article. The rule itself will be described in detail in a later chapter.

5. The Mannheim Slide Rule.— Amédée Mannheim, a Frenchman with a German name, was born in 1831 and died in 1906 after a long and illustrious career as an artillery officer and as Professor of Mathematics at L'École Polytechnique in Paris. It was in 1850 when only about nineteen years old that he designed his slide rule. He was then an artillery officer at Metz, one of the great fortresses of Alsace-Lorraine, where he had gone as Lieutenant of Artillery after he entered L'École Polytechnique in 1848. Besides his military contributions and his slide rule he has contributed largely by his researches to geometry and mechanics, and as Colonel Mannheim homage was rendered him at L'École Polytechnique on the occasion of the celebration of his seventieth birthday in 1901.

Lieutenant Mannheim designed his slide rule in 1850 but it was not publicly announced until December, 1851, when he described the rule and its use in a publication entitled "Modified Calculating Rule.—Instructions." The Mannheim Rule was adopted by the French Artillery and manufactured by the famous firm of Tavernier-Gravet. It was several years, however, before it became known in other countries and it did not come into general use until a book on the slide rule which highly recommended the Mannheim Rule was published in 1859 by an Italian named Q. Sella. This book made such an impression that the rule was soon in general use in Europe and the year 1859 is sometimes erroneously given as the year of its invention. The Mannheim Rule was hardly known in England for twenty years after its invention and was very little used in the United States before 1890.

6. Recent Developments. — "Great activity has been shown in recent times in the direction of perfecting the mechanical execution of slide rules. In this, Germany took the lead. After the slide rules used in Germany had been imported from France for a period of 20 or 30 years German manufacturers gained the ascendancy. In 1886 Dennert & Pape in Altona began to mount the scales of numbers upon white celluloid instead of boxwood or, as in some cases, metal. This method has now become well nigh universal. In America celluloid scale rules were being sold by Keuffel & Esser as early as 1886 and since 1886 this firm has manufactured them in Hoboken, N. J. A prominent German firm engaged in the manufacture of slide rules is that of Albert Nestler in Baden. The firm of A. W. Faber has factories in Germany and branch houses in England, France and the United States.

"Slide rule scales are now usually engine divided and have reached a very high state of perfection in both workmanship and mathematical accuracy.

"For the purpose of securing greater accuracy in the use of the slide rule without increasing the length of the divisions, to secure, say, the fourth [or fifth] figure with a 25-centimeter rule, attachments have been made which [are either magnifying glass runners or verniers and frequently] go under the name of "cursors." An early attempt along this line was a vernier applied by J. F.

Artur in Paris. Similar suggestions were made by O. Seyffert in Germany. Perhaps the best known of these is the Goulding Cursor, which allows the space between two consecutive smallest divisions to be divided into ten parts."[1] Probably the most convenient magnifying runner is that supplied by Keuffel & Esser, although very similar runners are supplied by other manufacturers. This is mounted in a metal frame and is applied to the rule by fastening it to the regular runner with a spring clamp. This firm also supplies a so-called "decimal indicator" which is a small semicircular scale attached as an integral part of the runner, and bearing divisions indicated by a pointer as subdivisions of the main scale.

Many special forms of slide rule have been developed for particular uses; some of these will be described in a later chapter. In his "History of the Slide Rule" Cajori lists 256 slide rules which in 1909 had appeared since 1800 and says that the list is probably not at all complete.

7. The Slide Rule in the United States. — Before about 1880 the slide rule was but little known and very little used in the British American Colonies and the United States and references to it in American technical and mathematical literature were few and sketchy. An account of the rule which probably reached the engineering

[1] From Cajori's "History of the Slide Rule."

profession more widely than any other was published in New York in 1856 in the "Mechanics and Engineers Reference Book." This book was written by Charles Haslett, Civil Engineer, and Charles W. Hackley, Professor of Mathematics at Columbia University (then Columbia College), and devotes five pages to the slide rule.

One of the earliest slide rules produced in America was Palmer's "Computing Scale" which appeared in Boston in 1844. It was a circular rule eight inches in diameter, thus providing a scale about twenty-five inches long. It was later modified and carried all the usual scales of that time but seems not to have been used outside of Massachusetts and New York.

The slide rule first began to come into real use in the United States soon after 1880 when the Mannheim Rule was first introduced. About that time it was made a part of the required courses in mathematics and engineering at Washington University in St. Louis, an institution famous for its early work in mathematics, astronomy, physics and engineering and for long the scene of the labors of William Chauvenet, the celebrated American mathematician, astronomer and teacher. In 1881 the well known Thacher Cylindrical Slide Rule was patented by Edwin Thacher, a graduate of Rensselaer Polytechnic Institute and a bridge engineer. This rule will be described in a later chapter. Following this invention, the work at Washington University, and the introduction of the Mannheim Rule, the use of the slide rule began to spread but the rule was not generally popular until William Cox began his work in 1890.

In that year Cox began a long campaign of propaganda and education in the columns of the "Engineering News," published in New York. He took the Mannheim Slide Rule as standard and wrote articles describing the rule and its use and preaching its value and importance to engineers and other technical workers. He patented several modifications of the Mannheim Rule which were produced by the firm of Keuffel & Esser and in 1891 wrote a manual of instruction for users of the slide rule, which was also published by Keuffel & Esser and which, with revisions, remained in wide use for thirty-five years.

In 1891 Cox patented his duplex slide rule. In this rule there are three strips of equal thickness, the slide and two outer pieces, these outer pieces being rigidly held together by metal cleats and grooved on the inner edges so that the slide fits snugly and slides freely between them with the faces of all three strips flush on both sides. This rule is thus mechanically the same as the Seth Partridge rule of 1657. The novelty and advantage of the Cox Duplex rule lay in its use of the Mannheim scales and the printing of scales on both sides of the entire rule. The inverted, equal parts, cube and folded scales were, about 1897, added to the Cox Duplex rule and the resulting rule called the "Polyphase Duplex" rule. This rule has now superseded the original Cox duplex and is widely used among engineers.

The standard Mannheim Rule carries scales A and D on the stock or body of the rule and scales B and C on the face of the slide, with scales of sines, tangents and equal parts on the back of the slide. In 1900 Keuffel & Esser added an inverted C scale to the face of the slide and a scale of cubes of C scale numbers to the stock. This modification of the Mannheim Rule is called the "Polyphase Mannheim" rule. In the same year this same firm also brought out an improvement in the mechanical construction of the Mannheim Rule. This consists in making one side of the groove in which the slide fits, movable on the main body of the stock. With this improvement the pressure of the sides of the groove on the slide can be adjusted to supply the proper friction for the best operation of the slide. The duplex rules of Keuffel & Esser are also made adjustable by making one of the strips of the stock, between which the slide moves, moveable perpendicularly to its length between the metal cleats which hold them.

The most complete of the standard form rules, also supplied by Keuffel & Esser, is their "Log-log Duplex" rule, brought out in 1908. It is constructed exactly like the Polyphase Duplex with the exception that it is wider, and in addition to all the scales of that rule carries a log-log scale in four parts, each running the full length of the rule. One gives $x = e^n$ when x is between 0 and 1 (that is, a fraction) and the other three give $x = e^n$ when x is between 1 and 22000. This improvement adds greatly to the value of the rule in connection with the calculations of modern physics and engineering.

Keuffel & Esser supply a number of other types of slide rule for special purposes, some of which will be described in a later chapter. Besides this firm whose series is perhaps the most complete, there is another American firm which produces excellent slide rules; this is Eugene Dietzgen Company. Their rules are of the forms now standard in the United States and a large part of the world, but those rules other than the standard Mannheim Rule bear trade names specially adopted by the producer. Later descriptions of standard forms of slide rules in this book will apply equally well to the rules of all makers who supply such rules. The types of rules described in this book will be those most used in the United States and the names used in cases where the rule is supplied by only one maker will be those adopted by that maker.

CHAPTER II

THEORY AND OPERATION OF THE MANNHEIM SLIDE RULE

A. THEORY OF THE RULE.

8. Some Properties of Exponents. — We give here for reference and review a brief summary of some of the chief properties of algebraic exponents. Their demonstrations can be found in any good text book of algebra and the statements given here are to be considered as more descriptive than demonstrative.

If a, b, m, n are any real numbers (that is, numbers not involving $\sqrt{-1}$), then a^m represents the continued product of a taken m times as a factor and

$$a^m \times a^n = a^{m+n} \qquad (1)$$

represents the product of a used $m + n$ times as a factor. Similarly

$$a^m \div a^n = a^{m-n} \qquad (2)$$

indicates the use of a as a multiplier m times followed by its use as a divisor n times so that the net result is the use of a as a factor $m - n$ times. In the same way as in (1) if a is used m times as a factor and the result so obtained is itself in turn used n times, then a will be used altogether $m \times n$ times. Therefore

$$(a^m)^n = a^{mn}. \tag{3}$$

If in (2) we make $n = m$ we get

$$a^m \div a^m = a^{m-m} = a^0;$$

but $a^m \div a^m = 1$ and therefore

$$a^0 = 1. \tag{4}$$

Therefore if in (2) we put $m = 0$ we get $a^0 \div a^n = a^{0-n} = a^{-n}$; but $a^0 \div a^n = 1 \div a^n$. Therefore

$$a^{-n} = \frac{1}{a^n}. \tag{5}$$

Similarly $$\frac{1}{a^{-n}} = a^n. \tag{6}$$

By definition the square root of a number is that number which multiplied by itself produces the given number. Thus $\sqrt{a} \times \sqrt{a} = a$. Similarly $\sqrt[3]{a} \times \sqrt[3]{a} \times \sqrt[3]{a} = a$, for the cube root, and for the nth root $\sqrt[n]{a} \times \sqrt[n]{a} \times \sqrt[n]{a} \times \ldots$ n times $= a$. That is, $\sqrt[n]{a}$ raised to the nth power gives a, or $(\sqrt[n]{a})^n = a$. But we have also by (3), $(a^{1/n})^n = a^{1/n \times n} = a^1 = a$. Hence

$$a^{1/n} = \sqrt[n]{a}. \tag{7}$$

Again by (3), $(a^{1/n})^m = a^{(1/n \times m)} = a^{\frac{m}{n}}$. \hfill (8)

Also, we get the same result if we raise a to the mth power and then take the nth root of the quantity so obtained or if we first take the nth root of a and then raise this quantity to the mth power; that is $\sqrt[n]{(a^m)}$ $=(\sqrt[n]{a})^m$. But by (7) $\sqrt[n]{a} = a^{1/n}$, and hence according to (8), $(\sqrt[n]{a})^m = (a^{1/n})^m = a^{m/n}$. Thus we have

$$a^{m/n} = \sqrt[n]{(a^m)}. \tag{9}$$

Now by definition $(ab)^2 = (ab) \times (ab) = (aa) \times (bb) = a^2b^2$; $(ab)^3 = (ab) \times (ab) \times (ab) = (aaa) \times (bbb) = a^3b^3$; and so on for any other exponent. Thus we can write

$$(ab)^m = a^m b^m. \tag{10}$$

From this result we can write $(ab)^{1/n} = a^{1/n}b^{1/n}$, or by (7)

$$\sqrt[n]{ab} = \sqrt[n]{a} \times \sqrt[n]{b}. \tag{11}$$

In the same way $\left(\dfrac{a}{b}\right)^2 = \dfrac{a}{b} \cdot \dfrac{a}{b} = \dfrac{aa}{bb} = \dfrac{a^2}{b^2}$, and so on; thus

$$\left(\frac{a}{b}\right)^m = \frac{a^m}{b^m} \tag{12}$$

and
$$\sqrt[n]{\frac{a}{b}} = \frac{\sqrt[n]{a}}{\sqrt[n]{b}}. \tag{13}$$

(It is a common error for students to write, analogously to (12) and (13), $\sqrt{(a+b)} = \sqrt{a} + \sqrt{b}$, $\sqrt{(a-b)} = \sqrt{a} - \sqrt{b}$, and similarly for other roots and powers, but that this is not true is easily seen by taking numerical values for a and b. Therefore we can NOT write $\sqrt[n]{a \pm b} = \sqrt[n]{a} \pm \sqrt[n]{b}$, $(a \pm b)^n = a^n \pm b^n$.)

9. Logarithms. — Based on the formulas of Article 8 we give here in the same summary manner some of the chief elementary properties of logarithms. By definition "the logarithm of any number A to the base b is the exponent m of the power to which b must be raised in order to produce A." Thus if

$$\left. \begin{array}{r} m = \log_b A \\ b^m = A. \end{array} \right\} \quad (14)$$

then

and a logarithm obeys all the rules of exponents. Thus if $m = \log_b A$ and $n = \log_b B$, then by the second of (14) and by (1) $AB = b^m \times b^n = b^{(m+n)}$ and since now $b^{(m+n)} = AB$, then by (14) $\log(AB) = m + n$. That is,

$$\log(AB) = \log A + \log B. \quad (15)$$

In a similar manner from (2) and (14) we get

$$\log\left(\frac{A}{B}\right) = \log A - \log B. \quad (16)$$

From (15) $\log(A^2) = \log(AA) = \log A + \log A$, that is $\log(A^2) = 2(\log A)$.

Similarly $\log(A^3) = \log(AAA) = \log A + \log A + \log A = 3(\log A)$, and so on for any exponent. Thus

$$\log(A^m) = m \log A. \quad (17)$$

If instead of the exponent m we use the exponent $\frac{1}{n}$ this equation gives $\log(A^{1/n}) = \frac{1}{n}(\log A)$. But by (7) $A^{1/n} = \sqrt[n]{A}$. Therefore

$$\log(\sqrt[n]{A}) = \frac{\log A}{n}. \quad (18)$$

If now we use (17), $\log[(AB)^m] = m\log(AB)$.
Hence by (15)
$$\log[(AB)^m] = m(\log A + \log B). \tag{19}$$
Similarly using (18) and (15),
$$\log[\sqrt[n]{AB}] = \frac{1}{n}(\log A + \log B); \tag{20}$$
using (17) and (16),
$$\log\left[\left(\frac{A}{B}\right)^m\right] = m(\log A - \log B); \tag{21}$$
using (18) and (16),
$$\log\left[\sqrt[n]{\frac{A}{B}}\right] = \frac{1}{n}(\log A - \log B), \tag{22}$$
and so on for any combination of the formulas (1) to (13). In particular we will note that since by (4) $b^0 = 1$, then by (14) to any base whatever

and since $b^1 = b$,
$$\left.\begin{array}{r}\log 1 = 0 \\ \log_b b = 1.\end{array}\right\} \tag{23}$$

Certain of the formulas in this article are fundamental in the theory of the slide rule; all are useful toward an understanding of its operation.

10. Principle of the Slide Rule. — The slide rule is a device by means of which multiplication, division, raising to powers, root extraction and their combinations may quickly and easily be performed by carrying out mechanically the logarithmic operations indicated by formulas (15) to (18) and the related formulas. This is readily understood by reference to the following diagrams.

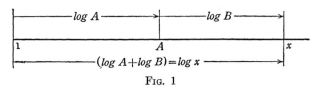

Fig. 1

Suppose we wish to multiply A by B and let $x = AB$. Then by formula (15) $\log x = \log A + \log B$. We look up $\log A$ in a table of logarithms and lay it off to any desired scale of drawing on a straight line. Let this be the length from 1 to A in Fig. 1. To the same scale we lay off the length A to x for $\log B$. Then the sum of these two lengths, 1 to x on the line, is $(\log A + \log B)$ and by the formula this equals $\log x$. So we measure the total length 1 to x on our line and determine the number x to which this length corresponds, by using the drawing scale. This number is the product of A and B and we have thus found $x = AB$ mechanically and without any multiplying.

The logarithmic line begins with 1 instead of 0 (zero) because its length represents the logarithms of numbers while the figures we have marked on the line are the numbers themselves which we are multiplying, and by formula (23) the zero length represents the logarithm of 1.

Using the same sort of scale we find $y = A \div B$ as indicated in Fig. 2. By formula (16) we know that when $y = A \div B$, then $\log y = \log A - \log B$. Using our scale line as before we lay off from 1 the length cor-

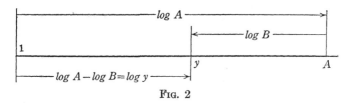

Fig. 2

responding to $\log A$, bringing us to the point A on the scale. From this length we subtract the length $\log B$ by starting at A and laying off $\log B$ in the reverse direction, bringing us to the point y. Since the length 1 to y is now equal to $(\log A - \log B)$ it represents to our chosen scale the logarithm of y. From the drawing scale we then determine the number corresponding to this length and we have the quotient of $A \div B$.

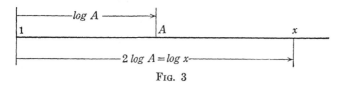

Fig. 3

Suppose now that we wish to find the square of some number A on our scale. By formula (17) we know that $\log(A^2)$ is 2 times $\log A$, so we need only lay off on our line twice the scale length corresponding to A and see what number it reaches on the line. This is shown in Fig. 3. Here the number $x = A^2$ is found from the length of $\log x$ (1 to x on the line) as explained above. In the same way we could find A^3 by putting $x = A^3$ and finding the length corresponding to $\log x = 3 \log A$. Applying formula (18) to our scale line we can easily find $y = \sqrt{A}$ by laying off as in Fig. 4 the length $\log A$

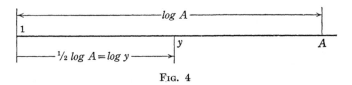

Fig. 4

and then finding the scale number corresponding to $\frac{1}{2} \log A$. By the same method we could just as easily find $y = \sqrt[3]{A}$.

A more rapid method of finding squares and cubes and roots is based on the following considerations: When drawn to the same scale, a line of a certain length which is twice a shorter length represents the logarithm of the square of the number whose logarithm is represented by the shorter length. Therefore when we use two such drawing scales on the same line, or on two lines side by side, a length on the line drawn to the smaller scale will represent the square of a number represented by the same length on the large scale line. On the other hand a length on the large scale line will represent the square root of the number represented by the same length on the small scale line.

Thus let the two lines of Fig. 4 be drawn separately

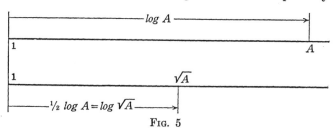

Fig. 5

from the same starting point, as in Fig. 5. Now let the scale of the upper line be reduced so that the length representing $\log A$ on it equals the length representing $\log \sqrt{A}$ on the other scale. This gives us Fig. 6. By

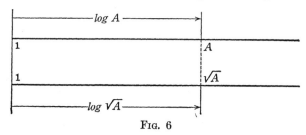

Fig. 6

drawing two such scale lines therefore, and laying a straight edge across them at right angles at a point on the upper line corresponding to any chosen number it will cross the lower line at a point corresponding to the square root of the number. By choosing a number on the lower scale line its square is similarly found on the upper line.

The processes described in connection with Figs. 1, 2, 6 embody the principles of the slide rule and have here been described and explained at length because of their extreme importance.

11. Application of the Slide Rule Principle. — The details of the operations described and explained in the preceding article are somewhat laborious and their performance is much facilitated if instead of drawing one

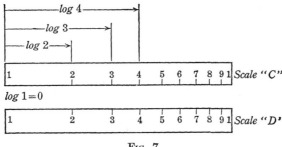

Fig. 7

scale line as in Figs. 1 and 2 and separately measuring off the proper lengths on it, we draw two such lines to the same scale on two strips of wood, laying off to a convenient scale the *lengths* corresponding to the *logarithms* of the numbers, and marking on the lines the FIGURES representing the NUMBERS themselves. Such an arrangement is represented in Fig. 7.

APPLICATION OF THE SLIDE RULE PRINCIPLE 31

The scale divisions are not uniform because when the logarithms are in arithmetical progression the numbers are in geometrical progression as in the tables below; and when the numbers are in arithmetical progression the logarithms are in geometrical progression as on the scales in Fig. 7.

Number   = 1, 10, 100, 1000, 10000, 100000, 1000000,
Logarithm = 0, 1,   2,   3,   4,   5,   6,
Base   10 10   10   10   10   10   10.

Number   = 1, 2, 4, 8, 16, 32, 64, 128, 256, 512,
Logarithm = 0, 1, 2, 3, 4, 5, 6,   7,   8,   9,
Base   2 2 2 2   2   2   2   2   2   2.

The use of the scales of Fig. 7 in multiplication is shown in Fig. 8 where 2 is multiplied by 3. Thus $2 \times 3 = 6$ is

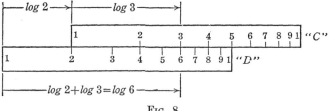

Fig. 8

found by simply placing the beginning of the "C" scale at the multiplicand 2 on the "D" scale and at the multiplier 3 on the "C" scale reading the product 6 on the "D" scale.

The operation described in connection with Fig. 2 is now carried out very easily by placing the two wooden scales together as in Fig. 9, which shows the division of 8 by 4. Here the divisor 4 on the "C" scale is placed at the dividend 8 on "D" and the quotient 2 is read on "D" at the 1 of "C."

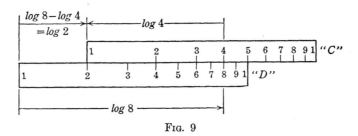

Fig. 9

As the same logarithm table serves for all numbers by shifting the decimal points or changing the characteristics, so in both Figs. 8 and 9 we could, by shifting the decimal points, read 2, 3, 6; 2, 30, 60; 20, 30, 600; .2, .3, .06; .2, .03, .006; or 8, 4, 2; 80, 40, 2; 0.08, .4, .2, etc., as might be necessary. This procedure will be explained in detail later. Also by making the scales of any desired length and subdividing the main divisions decimally, any numbers, whole, fractional or mixed, may be read as closely as desired.

The application of Fig. 6 to the wooden scale or rule is shown in Fig. 10. Here the "D" scale is the same as in Figs. 7, 8, 9 but "A," like the upper line in Fig. 6, is laid off to a scale half as great. The straight edge laid across the two scales at 9 on "A" crosses "D" at $3 = \sqrt{9}$. Conversely, in line with 6 on "D" is found $6^2 = 36$ on "A."

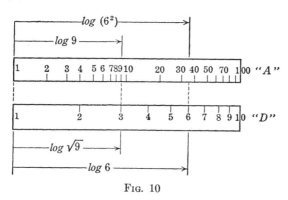

Fig. 10

The scales "A," "C" and "D" are incorporated in the Mannheim Slide Rule as described in the next article.

12. The Standard Mannheim Slide Rule. — The American standard slide rule is that designed by Lieutenant Mannheim and consists of a strip of wood about $10\frac{5}{8} \times 1\frac{1}{8} \times \frac{3}{8}$ inches ($27 \times 2.9 \times 1$ cm.) with a groove cut in one side about $\frac{9}{16} \times \frac{3}{16}$ inches (about 1.4×0.5 cm.) in which slides, with outer faces flush, a second strip of wood about $10\frac{5}{8} \times \frac{9}{16} \times \frac{1}{8}$ inches ($27 \times 1.4 \times 0.4$ cm.), guided by tongues fitting in smaller grooves in the edges of the larger groove. This arrangement is shown in section (not to scale) in Fig. 11 (b) and in plan (not to scale), with the sliding strip partly drawn out to the right, in Fig. 11 (c). The grooved part is called the "stock" or "rule" and the sliding strip the "slide." On the upper face of each are two divided scales $9\frac{27}{32}$ inches (25 cm.) long, arranged as shown in Fig. 12 and marked on celluloid or ivory facings. These are the A, C and D scales already described and a fourth scale called the "B" scale.

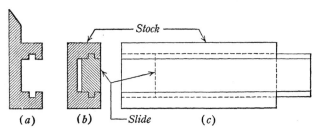

Fig. 11

Fig. 12

The A and B scales are exactly alike and the same as the A scale in Fig. 10. The C and D scales are exactly alike and the same as in Figs. 8, 9. On the opposite side of the slide are three scales marked "S," "T" and "L."

GRADUATION AND READING OF THE SCALES 35

These scales will be described later. A glass plate with a hair line on its under side which reaches across all the scales at right angles, is attached to the rule and arranged to slide along its full length. This is called the "runner" or "indicator." In most cases the stock bears an extension or "lip" on the side bearing the A scale as shown in section in Fig. 11 (a). This lip carries on its beveled edge an ordinary 10 inch or 25 centimeter rule marked on celluloid or ivory. A complete rule as manufactured by Keuffel & Esser is shown in detail and to scale in Fig. 13 with the B and C scales face upward on the slide.

FIG. 13. MANNHEIM SLIDE RULE

Rules with 5, 8, 16, and 20 inch scales are frequently used but the 10 inch (25 cm.) length is the one most used.

B. OPERATION OF THE RULE.

13. Graduation and Reading of the Scales. — As shown plainly in Fig. 13 the C and D scales are alike and divided into 10 main parts, reading 1, 2, 3, 4, 5, 6, 7, 8, 9, 1. These are generally referred to as 1 to 10 although, since the graduation is decimal, they may be taken as .1 to 1, 1 to 10, 10 to 100. etc. The figure 1 at the left is called the "left index" and that at the right the "right index."

Each of these main divisions is divided into 10 parts, and each of these is in turn subdivided, though not all alike. Thus between 1 and 2 the first subdivisions are divided into 10 parts, between 2 and 4 into 5 parts, and between 4 and 10 into 2 parts. Between 1 and 2 the first subdivisions are themselves numbered 1 to 9 but the others are not usually numbered.

When the main figures are taken as units the first subdivisions are tenths, and between 1 and 2 the second subdivisions are hundredths, between 2 and 4 they are two hundredths, and between 4 and 10, five hundredths. By means of the hair line marker on the runner these smallest subdivisions may themselves be divided and the value of the divisions estimated by the eye. In this way between 1 and 2 ten-thousandths may be estimated directly (the fourth figure); between 2 and 4 hundredths may be estimated directly and with practice five thousandths (the third and sometimes fourth figure); between 4 and 10 hundredths may be estimated directly and in some cases, as between the tenth mark and the five hundredths mark, we may read .025 or .075 giving the third and in special cases the fourth figure. In general we say that on a 10 inch rule we read the fourth figure at the left and the third figure toward the right.

GRADUATION AND READING OF THE SCALES   37

As an example in scale reading consider the setting of the rule shown in Fig. 13. The left index of the C scale is midway between 119 and 12 or 120 on D; this is read 1195, "one one nine five," without reference to units or decimal point. Depending on the values and magnitude of the numbers we may be using in a particular case, this may be 1.195, 11.95, 119.5, 1195, 11950, etc., or .1195, .01195, .001195, etc.

Again, the hair line of the runner or indicator crosses the C scale at 346, "three four six." It crosses the D scale between 41 and 415 at a little more than three-fifths of the distance from 41, that is, to three figures, 413, "four one three." Similarly the right index of the D scale is just at the right of 835, that is, at 836, "eight three six."

Comparing the A and D or the B and C scales it is seen that A consists of two duplicate scales each of which runs 1 to 10, or considered as one scale it has three indexes and runs 1 to 10 to 100. Each numbered division is divided into 10 parts and from 1 to 5 these are subdivided, between 1 and 2 into 5 parts and between 2 and 5 into 2 parts. Thus between 1 and 2 the smallest subdivisions are two hundredths, between 2 and 5 they are five hundredths, and between 5 and 10, tenths. Thus with the indicator we can estimate hundredths, though not so accurately from about 8 to 10. With reference to the values of these numbers and the position of the decimal point the same thing applies to the A and B scales as to C and D. Thus if from the left index of A to the middle index the figures are taken as units, then from the middle index to the right the figures indicate tens and we read 10 to 100. In the same way as for C and D we may also read A and B as 10 to 100 to 1000, etc., or .001 to .01 to .1, .01 to .1 to 1, .1 to 1 to 10, etc. The position of the decimal point on the A and B scales and the relation of these scales to C and D will be discussed in detail farther on in this book and we shall then see that when the units of C or D are chosen then for certain purposes the units of B or A are fixed automatically.

The left hand or first half of the A scale is called A1 and the right hand or second half is called A2. Similarly the left and right halves of B are called B1 and B2. In what follows we shall use these designations.

As examples in reading the A and B scales consider again the "settings" (positions of slide and runner) of the rule in Fig. 13. The left index of B is midway between 142 and 144, at 143, on A. The middle index of A is at 700 on B. The middle index of B is at 143 on A, and the right A index is at 700 on B. The indicator is at 171 on A and at 120 on B.

Just to the right of 3 on A1 and B1 and to the left of 8 on A2 and B2 are two special marks called "gauge points."

GRADUATION AND READING OF THE SCALES

These are respectively $\pi = 3.1416$ and $0.7854 = \frac{1}{4}\pi$. Their uses will be explained later.

The descriptions of the scales and reading are given here in detail and they should be carefully studied and practice settings made and read until they are thoroughly mastered. When this is done statements of complicated settings and operations are easily understood and followed without the necessity of detailed descriptions or elaborate diagrams. Thus the setting of the slide and runner may be stated as follows: "Left C index to 1195 on D, Runner to 346 on C, Read 413 on D"; or "Left B index to 143 on A, Runner to 120 on B, Read 171 on A." These may be diagrammed separately as in Fig. 14 (a) and (b), or in one combination as in Fig. 14 (c).

| A | To 143 | Read 171 |
| B | Set 1 | R to 120 |

(a)

| C | Set 1 | R to 346 |
| D | To 1195 | Read 413 |

(b)

A		Read 171
B		
C	Set 1	R to 346
D	To 1195	

(c)

Fig. 14

Forms of statements of settings similar to those given in quotation marks above, or setting diagrams such as those of Fig. 14, will be much used in what follows.

14. Multiplication with the Slide Rule. — The C and D scales of the Mannheim Slide Rule are used in multiplication exactly as indicated with the two detached scales C, D in Fig. 8 but with the actual rule the slide is held in place in the groove of the stock by friction and it is not necessary to press the scales together by hand; also the indicator is a great aid in making accurate settings and readings.

Suppose we wish to multiply 2 by 3. We say "To 2 on D set left 1 of C; Runner to 3 on C; At indicator read 6 on D." It is to be noticed that the numbers are taken in the order in which they appear in the multiplication as written:

$$2 \times 3 = 6$$
$$(1)\ (2)\ \ (3)$$

Except in special cases multiplication is always performed on the C, D scales and the operation is always carried out as follows

(1) To *Multiplicand* on D set C index.
(2) To *Multiplier* on C set runner indicator.
(3) At indicator on D read *Product*.

The operation $2 \times 3 = 6$ is shown in detail in Fig. 8 and is diagrammed in Fig. 15 (a). Fig. 15 (b) shows multiplication in general.

C	Set 1	R to 3
D	To 2	Read 6

(a)

C	Set 1	R to Multiplier
D	To Multiplicand	Read Product

(b)

Fig. 15

In the case of $2 \times 3 = 6$ the readings are exact and simple. Suppose, however, we wish to multiply 1.195 by 3.46 and read the product correct to the nearest three figures. The operation is:

(1) To 1.195 on D set C index.
(2) To 3.46 on C set runner.
(3) At indicator on D read 4.13.

To simply set one index of C at 1.195 on D is tedious and also liable to error since the black line at 1 has some width. But if we first set the very fine hair line of the indicator at 1.195 on D and then bring up the slide so that the 1 line is exactly under the indicator hair line the setting is generally much more easily and accurately made. Similarly if we have to locate the point 3.46 on C with the unaided eye and then determine the D scale reading which is in line with this, as is done in Fig. 8 with the 3 and 6, there is not only difficulty in reading but liability to error. It is thus seen that the indicator is at the same time a labor and time saver and also when properly used an insurance against many errors. We shall see later that the indicator is also a necessity in certain forms of settings.

The multiplication $1.195 \times 3.46 = 4.13$ is already set and described in Fig. 13 and diagrammed on C, D in Fig. 14 (b). There we read only the figures 1195, 346, 413. The values of the numbers are determined by the position of the decimal point and when it is once placed in the multiplicand and multiplier its position in the product is determined by those. Thus we know that 1.195×3.46 cannot be 0.413, neither can it be 41.3. A number slightly greater than 1 multiplied by one nearly equal to $3\frac{1}{2}$ must give a product closely equal to 4. So the product can only be 4.13. Again, if our multiplication were 11.95×34.6 it would be nearly 12×35 and the product must be in the neighborhood of 400, that is, 413. So in any case a knowledge of the values of the factors gives at once the approximate value of the product and when the scale reading gives the figures the proper position of the decimal point is known at once. We shall later give definite rules for locating decimal points, after we have described the operation of the rule, but these should not be used blindly. The results should always be checked by a knowledge of simple arithmetic. As a matter of fact physicists and engineers almost always locate decimal points in slide rule results by inspection.

Another question now arises. If we carry out the multiplication 1.195×3.46 in the ordinary way we get 4.13470. To the nearest three figures this would be written 4.13; thus the slide rule automatically makes our approximations for us. If the objection is made that the result is not exactly correct it must of course be agreed that it is not, but decimal fractions rarely are exact and the conditions of the problem in hand or simply choice will usually determine just how far to carry the approximation to exactness. In the present case the error is $413470 - 413000 = 470$ parts in 413470, that is $\frac{470}{413470}$ $= .001137 = .1137\%$, about $\frac{1}{9}$ of 1% or 1 part in 900. In such operations as those involving counting where figure accuracy is required the slide rule can only be used as an approximate check, but in measurement and in general engineering work where proportional or percentage accuracy is considered, the slide rule is in general sufficiently accurate, reading results to about one tenth of one per cent.

We now consider a slightly different type of multipli-

cation operation. Let us multiply 43.3 by 5.25. This is stated:
  (1) To 43.3 on D set C index.
  (2) To 5.25 on C set runner.
  (3) At indicator on D read *Product*.

We readily set the indicator line at 433 on D and as usual bring up the left C index line to the indicator line and step (1) is complete. When we attempt to bring the runner (indicator) to 525 on C, however, we find this to be impossible; that point is beyond the end of D. As we ordinarily say, it "runs off the scale." We get around this difficulty by the following considerations: Since the logarithmic scale is decimal we may read it either way, that is, call the left index 1 and the right 10 or the right index 1 and the left .1. Thus since we may take either index as 1 we may use either index to set on the multiplicand in multiplication. In the present case then let us try the right C index at 433 on D; we now easily bring the runner to 525 on C. Step (3) of the operation then gives for the product the figures 227. Now $43\frac{1}{3} \times 5\frac{1}{4}$ cannot be 22.7 or 2270; the result is 227. The diagram

C	Set 1	R to 525
D	To 433	Read 227

(a)

A	To 433	Read 227
B	Set 1	R to 525

(b)

Fig. 16

MULTIPLICATION WITH THE SLIDE 'RULE

of this operation is made in the usual manner in Fig. 16 (a) with "C index" here indicating the right hand index.

Multiplication may be carried out with the A and B scales in precisely the same manner as with C, D, the B scale (on the slide) being used instead of C and the A scale (on the stock) instead of D. Either half of A or B may be used. Thus to apply to either set of scales the statement corresponding to Fig. 15 (b) might read:

(1) To *Multiplicand* on STOCK set SLIDE index.
(2) To *Multiplier* on SLIDE set runner indicator.
(3) At indicator on STOCK read *Product*.

The last multiplication, 43.3 × 5.25 = 227, would thus be carried out on A, B as indicated in Fig. 16 (b). The A, B scales are used for multiplication in certain combination settings in connection with C, D but for straight multiplication, C, D should always be used.

C	Set 1	R to 47.6
D	To 73.5	Read 3500

(a)

C	Set 1	R to 6.78
D	To .1473	Read 1

(b)

FIG. 17

Two sample multiplications are given in Fig. 17 and these, together with those already discussed, and many others, should be practiced until the operations are carried out without hesitation.

(a) 73.5 × 47.6 = 3500   (b) .1473 × 6.78 = 1.000

THEORY AND OPERATION

In order to carry out such a multiplication as $1.5 \times 3 \times 2 = 9$ we may first multiply $1.5 \times 3 = 4.5$ in the usual manner as in Fig. 18 (a) and then multiply $4.5 \times 2 = 9$ as in Fig. 18 (b). It will be noted that when we have read the 4.5 in (a) we ordinarily leave the runner unmoved on 4.5, for 4.5 is now the multiplicand of operation (b) and we ordinarily mark the multiplicand with the runner so as to bring up the slide to set the index accurately. Since we are not to move the runner from the 4.5 then until the slide is reset, we do not need to read the first product 4.5 but may combine the double operation into one double setting and read only the final result. This operation is shown in Fig. 18 (c).

C	Set 1	R to 3
D	To 1.5	Read 4.5

(a)

C	Set 1	R to 2
D	To 4.5	Read 9

(b)

C	Set 1	R to 3	1 to R	R to 2
D	To 1.5			Read 9

(c)

FIG. 18

MULTIPLICATION WITH THE SLIDE RULE   47

The double setting is stated as follows:
- (1) To 1.5 on D set C index.
- (2) To 3 on C set runner.
- (3) To runner set C index.
- (4) To 2 on C set runner.
- (5) At indicator on D read 9.

Similarly let us multiply $1.72 \times 2.94 \times 1.265$:
- (1) To 1.72 on D set C index.
- (2) To 2.94 on C set runner.
- (3) To runner set C index.
- (4) To 1.265 on C set runner.
- (5) At indicator on D read 6.40.

Here we did not read or need the intermediate product 5.06 of the first two factors.

In the same way any number of factors may be multiplied as follows. Let
$$a \times b \times c \times d = X$$
Then
- (1) To a on D set C index.
- (2) To b on C set runner.
- (3) To runner set C index.
- (4) To c on C set runner.
- (5) To runner set C index.
- (6) To d on C set runner.
- (7) At indicator on D read X.

It is to be noted that we read and set the multipliers on the slide only and alternately move the slide and runner, reading only the first factor and the final product on D. Thus we may state the above operation concisely as follows:

Set slide index to a on D, runner to b on slide, slide to runner, runner to c, slide to runner, runner to d, Read X on D;

and this series of operations is repeated for any number of factors $a \times b \times c \times d \times e \times \ldots = X$, alternately setting the runner to the factors (after the first) and the slide index to the runner.

15. Division and Reciprocals. — Division is the inverse of multiplication and the slide rule operation for division is exactly that described in connection with Fig. 9 or the inverse of that shown in Fig. 8. Thus we can consider that Fig. 8 shows either the operation $2 \times 3 = 6$ or $6 \div 3 = 2$ and by comparison of Figs. 8, 9 with the setting diagram Fig. 15 (a) we see that the diagrams for $6 \div 3 = 2$ and $8 \div 4 = 2$ are those of Fig. 19 above. The operation $8 \div 4 = 2$ may be stated as follows:

C	Set 3	Under 1
D	To 6	Read 2

(a)

C	Set 4	R to 1
D	To 8	Read 2

(b)

FIG. 19

(1) To 8 on D set 4 on C.
(2) To C index set runner.
(3) At indicator on D read 2.

If the A, B scales are used it is stated:

(1) To 8 on A set 4 on B.
(2) To B index set runner.
(3) At indicator on A read 2.

DIVISION AND RECIPROCALS

In general for either A, B or C, D:
(1) To *Dividend* on stock set *Divisor* on slide.
(2) To slide index set runner.
(3) At indicator on stock read *Quotient*.

It is to be noted that the operation is carried out in exactly the reverse of the order of multiplication, thus:

$$\text{Multiplicand} \times \text{Multiplier} = \text{Product}$$
$$\underset{(1)}{2} \times \underset{(2)}{3} = \underset{(3)}{6}$$

but

$$\text{Dividend} \div \text{Divisor} = \text{Quotient}$$
$$\underset{(1)}{6} \div \underset{(2)}{3} = \underset{(3)}{2}$$

It is now at once obvious that the setting of the rule in Fig. 13 may be viewed as that of $413 \div 346 = 1195$ (reading the figures only and ignoring decimal point), which would be stated and diagrammed as:

(1) To 413 on D set 346 on C.
(2) To C index set runner.
(3) At indicator on D read 1195.

C	*Set* 346	*R to* 1
D	*To* 413	*Read* 1195

Fig. 20

Of course Fig. 13 shows only the first step of the operation, setting 346 on C to 413 on D (by means of the runner). It is seen, however, that the C index is already at 1195 on D and when the runner is brought to 1 on C it will exactly indicate the figures 1195 on D.

50   THEORY AND OPERATION

The decimal point may now be determined by inspection. Thus if the numbers are 413 and 346 the quotient is obviously slightly greater than 1 and the decimal point is placed to give 1.195. If we are dividing 41.3 by 3.46 it is obvious that the quotient is very nearly $41 \div 3\frac{1}{2}$ or about 12, and when the decimal point is placed we have 11.95.

If Fig. 13 is carefully examined or if the setting of the rule as shown in Fig. 13 is reproduced on the actual slide rule it will be seen that the reading which we have called 413 is slightly greater than 413 but hardly 414; it should be read 4135. If the division $413.5 \div 346$ is carried out by long division the quotient does not come out exact, but to six figures it is 1.19364 and the slide rule error is $119500 - 119364 = 136$ parts in 119364, that is, $\frac{136}{119364}$ $= .001138 = .1138\%$, or about $\frac{1}{9}$ of one per cent.

As a final example let us divide 164 by 34. The statement is

(1) To 164 on D set 34 on C.
(2) To C index set runner.
(3) At indicator on D read *Quotient*.

Step (1) is easily and exactly made; when we attempt step (2), however, only the right C index can be reached. We try this index and the quotient read on D is 4.825, which is exactly correct. So in division there is only one choice of index and we do not have to reset the slide

DIVISION AND RECIPROCALS 51

to determine it. The setting for this division is Fig. 21. Incidentally, this setting furnishes an example of a four-figure scale reading toward the right end of the scale which is easily made. It also gives an example of a slide rule result which is exactly correct. Instead of having an error of about one tenth of one per cent the error is zero.

| C | Set 34 | R to 1 |
| D | To 164 | Read 4.825 |

FIG. 21

A *reciprocal* is a quotient with the dividend equal to 1. Thus the *reciprocal of a number* is a common fraction having the number as denominator and 1 as numerator and the slide rule furnishes a quick and direct reduction of such fractions to decimal fractions. Thus $\frac{1}{4} = 1 \div 4 = .25$ and this operation on the slide rule is stated in the usual manner as

(1) To 1 on D set 4 on C.
(2) To 1 on C set runner.
(3) At indicator on D read .25.

The diagram is Fig. 22 (a). From this we obviously have the following rule for reducing reciprocals to decimals, or for finding the reciprocal of any number which may itself be a whole number, decimal number or mixed number: "*To D index set the number on C and under C index read the reciprocal.*" The diagram is Fig. 22 (b).

52   THEORY AND OPERATION

The inverse operation, that of expressing any decimal fraction as a common fraction with the numerator 1, is sometimes useful. From the preceding rule and Fig. 22

C	Set 4	R to 1
D	To 1	Read .25

(a)

C	Set No.	Under 1
D	To 1	Read Recip.

(b)

Fig. 22

we have as the inverse rule: "*To the decimal on D set the C index and the number on C nearest the D index is the denominator.*" Thus for .0237 we find nearly $\frac{1}{42}$.

As a matter of fact, any common fraction may be considered as a division. Thus $\frac{3}{4} = 3 \div 4 = .75$ and we have at once the rule for reducing common fractions to decimals: "*To numerator on D set denominator on C and under C index find decimal.*" This rule is diagrammed in Fig. 23 (a).

The inverse of this rule is also sometimes useful, to express a decimal as a common fraction. Reversing the last rule we have: "*To the decimal on D set the C index and find on C and D the two simple numbers nearest to-*

gether; that on D is the numerator and that on C the denominator." This rule is diagrammed in Fig. 23 (b) A simple illustration is $.428 = \frac{3}{7}$, very nearly.

| C | Set Denominator | Under 1 |
| D | To Numerator | Read Decimal |

(a)

| C | Set 1 | Find Denominator |
| D | To Decimal | Over Numerator |

(b)

FIG. 23

16. The Decimal Point in Slide Rule Multiplication and Division. — We will now give the rules for locating decimal points which have been referred to in several places. This can best be done in connection with illustrative examples.

Thus consider the multiplication $24.3 \times 37 = 900$. We set the left C index to 24.3 on D and the slide projects from the rule toward the right. The same is the case with $1.88 \times 4.2 = 7.9$ and with $152.5 \times 58.2 = 8875$. Let us write these down together for comparison:

(1) $1.88 \times 4.2 = 7.9$
(2) $24.3 \times 37. = 900.$
(3) $152.5 \times 58.2 = 8875.$

In (1) the sum of the numbers of figures to the left of the decimal points in the factors is 2 and there is one figure to the left of the decimal point in the product. In (2) this sum is 4 and the product has three figures before the decimal point, and in (3) these figures are 5 and 4. In each case the number of figures before the decimal point in the product is one less than the sum of those in the factors. This will be found to be the case whenever the slide projects to the *right* in multiplication.

Consider the product $3.72 \times .242 = .9$; the same rule holds. In the product $.275 \times .32 = .088$ the corresponding sum for the factors is zero and not only are there no digits before the decimal point in the product, but also in one place after the decimal point there is no digit, the place being filled by a cipher (0). In this case the result of the examples (1), (2), (3) above would give $0 - 1 = -1$. Similarly when applied to the multiplication $.041 \times .222 = .0091$ we would have to write the rule as $-1 - 1 = -2$ and there are two places after the decimal point in the product filled by ciphers.

In order to state a rule for the decimal point in concise terms we define: The *characteristic* of a number is the number of digits before the decimal point; the characteristic of a decimal fraction is the number of ciphers immediately after the decimal point and is negative. (It is to be noted that the characteristic of a *number*

THE DECIMAL POINT 55

is not the same as the characteristic of the *logarithm* of a number.) For example, the characteristic of 1.88 and of 4.2 is 1; the characteristic of 24.3 and of 37. is 2; that of 152.5 is 3 and that of 8875. is 4. The characteristic of .9 is zero; that of .088 is − 1 and that of .0091 is − 2, etc. Using this definition we can now easily and concisely express the result of the several multiplications discussed above as a rule in the following form:

(i) RULE: *When the slide projects to the right in multiplication the characteristic of the product is 1 less than the sum of the characteristics of the factors.*

This rule holds for any number of factors by taking them two at a time; the detailed application to several factors will be given after the rules have been given for two factors and for division.

Consider now the operations:

(1) $73.5 \times 3.13 = 230.$ (3) $.675 \times .163 = .110$
(2) $56.7 \times .45 = 25.5$ (4) $8.33 \times .0021 = .0175$

In each of these the slide projects to the *left* and examination shows that the characteristic of the product is equal to the sum of the characteristics of the two factors. This gives the rule:

(ii) RULE: *When the slide projects to the left in multiplication the characteristic of the product equals the sum of the characteristics of the factors.*

Since the operation of division is the inverse of that of multiplication, the dividend taking the place of the product and the quotient and divisor those of the factors, the two rules (i) and (ii) when reversed give the rules for division. This will be made clearer by the following formulation: Let

M = characteristic of Multiplicand
m =   "   " Multiplier
P =   "   " Product
D =   "   " Dividend
d =   "   " Divisor
Q =   "   " Quotient,

then Rule (ii) gives for

Multiplication $\quad P = M + m$
Division $\qquad\quad D = Q + d$
therefore $\qquad\qquad Q = D - d$

and we have the following for division:

(iv) RULE: *When the slide projects to the left in division the characteristic of the quotient equals the characteristic of the dividend minus that of the divisor.*

Similarly Rule (i) gives for multiplication and division

$$P = (M + m) - 1$$
$$D = (Q + d) - 1$$
therefore $\quad Q = (D - d) + 1$

which gives the following:

(iii) RULE: *When the slide projects to the right in division the characteristic of the quotient equals the characteristic of the dividend minus that of the divisor, plus* 1.

THE DECIMAL POINT 57

The rules (i), (ii), (iii) and (iv) are combined into a simple chart below which is convenient for reference and easy to remember.

Characteristic of result	Slide LEFT	Slide RIGHT
Multiplication	Sum of Characteristics of 2 Factors	Sum $-$ 1
Division	Characteristic of Dividend $-$ that of Divisor	Difference $+$ 1

In order to see how these rules are applied in the multiplication of three or more factors consider the operation $4.2 \times 2.3 \times 19.3 = 186.3$. This may be considered as carried out in two steps as follows: $(4.2 \times 2.3) \times 19.3 = 186.3$. Referring to the chart above, since in the first setting the slide projects to the right we have as the characteristic of the product of the first two factors (though we do not read the product), $1 + 1 - 1 = 1$ and when this product is multiplied by the 19.3 the slide projects to the left and to the result 1 for the first step we add, according to the chart, the characteristic of 19.3, namely 2, and the final result is 3, which is the characteristic of the complete product 186.3. Again consider the operation $58.5 \times 4.65 \times 7.54 = 2050$. Here the slide projects to the left at each step of the operation and the characteristic of the product is the sum of the characteristics of all three factors. Comparing these two examples and trying out others of the same or greater number of factors it is seen at once that the decimal point is located in products of several factors in the following manner:

    (v) RULE: *To find the characteristic of the product of several factors add the characteristics of all the factors, and subtract 1 for each time that the slide projects to the right in carrying out the successive steps of the complete setting.*

The rules developed above for the location of decimal points in slide rule multiplication and division always give correct results but should not be used blindly or mechanically. With practice it will generally be found that decimal points can be located about as quickly and easily, and with greater appreciation of the significance of the results, by simple inspection. As mentioned once before, physicists, chemists and engineers generally use the method of inspection.

17. Combined Multiplication and Division. — After the operations of multiplication and division are mastered the combination of the two is simple. Thus consider the operation $(3 \div 2) \times 5 = 7.5$. The complete operation here consists of the two separate operations $3 \div 2 = 1.5$ and $1.5 \times 5 = 7.5$ and is diagrammed in two parts in Fig. 24.

Since in (b) we must set the C index to 1.5 on D and then bring the runner to 5 on the slide, and since at the

COMBINED MULTIPLICATION AND DIVISION   59

end of the operation (a) the C index is already at 1.5 on
D we need not stop to note the intermediate result 1.5

C	Set 2	R to 1
D	To 3	Read 1.5

(a)

C	Set 1	R to 5
D	To 1.5	Read 7.5

(b)

FIG. 24

but may combine (a) and (b) and obtain the final result
directly. This is done in Fig. 25 (a).

C	Set 2	R to 5
D	To 3	Read 7.5

(a)

C	Set 243	R to 38.4
D	To 19.45	Read 3.21

(b)

FIG. 25

As an example let us find the combined result (19.45 ÷ 243)
× 38.4 = 3.07. This is done in Fig. 25 (b). The result
is pointed off in accordance with the chart given in the

last article. Thus for the step 19.45 ÷ 243 the characteristic is 2 − 3 = − 1 and this used in the second part, (19.45 ÷ 243) × 38.4 gives − 1 + 2 = 1 as the characteristic of the final result, 3.07. These steps in the procedure of locating the decimal point may be combined into a single rule after the manner of Rule (v), as may also be done in the cases of multiple factors and divisors in the several cases discussed below, but the multiplication of rules tempts the student to memorize them and this is not a good plan. If rules are to be used, it is better to master the two rules given in the chart in the last article and when these are thoroughly understood to combine them in every case to suit the particular setting in hand.

The examples (3 ÷ 2) × 5 and (19.45 ÷ 243) × 38.4 may also be written as (3 × 5) ÷ 2 or (19.45 × 38.4) ÷ 243 but when carried out in this order the setting is more complicated and it is always better to perform the division first. For this purpose and for the purpose of extending the method to any number of factors and divisors it is better to write (3 ÷ 2) × 5 in the form $\frac{3 \times 5}{2} = 7.5$, or in general for any numbers whatever,

$$\frac{ac}{b} = X. \qquad (24)$$

This operation is diagrammed in Fig. 26 (a).

COMBINED MULTIPLICATION AND DIVISION

C	Set b	R to c
D	To a	Read X

(a)

C	Set b	R to c	d to R	R to e
D	To a			Read X

(b)

FIG. 26

The operation corresponding to formula (24) is stated as:

(1) To a on D set b on C.
(2) To c on C set runner.
(3) At indicator on D read X.

Comparing this with the statement of the operations of multiplication and division in Articles 14 and 15 it is seen that the combined operation requires no more steps than each of the separate operations. This advantage is even more apparent in the setting corresponding to the operation

$$\frac{ace}{bd} = X \qquad (25)$$

which is diagrammed in Fig. 26 (b) and stated as follows:

"To a on D set b on C, runner to c on slide, slide d to runner, runner to e, at runner on D read X."

By alternately moving the slide for divisors and the runner for multipliers this procedure applies to any number of quantities occurring in expressions like (25).

THEORY AND OPERATION

The decimal point in X is located by application of the rules of Article 16 as in the case of $\dfrac{19.45 \times 38.4}{243}$ above.

18. Proportion. — If we multiply both sides of the equation (24) by b we get

$$ac = bX \qquad (26)$$

and since this says that the product of a and c equals the product of b and X, a and c must be the means and b and X the extremes of a proportion, or vice versa. That is, either

$$b : a :: c : X \qquad \text{or} \qquad a : b :: X : c \qquad (27)$$

Thus the slide rule setting of Fig. 26 (a) will solve any proportion for any member X when the other three members are known. This is explained in the following paragraph.

The proportions (27) may also be written in the form $\dfrac{b}{a} = \dfrac{c}{X}, \dfrac{a}{b} = \dfrac{X}{c}$, or

$$\dfrac{b}{a} = \dfrac{c}{X}, \qquad \dfrac{a}{b} = \dfrac{X}{c}. \qquad (28)$$

Now these expressions are simply transformations of (26); other transformations are

$$\dfrac{X}{a} = \dfrac{c}{b}, \qquad \dfrac{c}{X} = \dfrac{b}{a} \qquad (29)$$

SQUARES, CUBES AND ROOTS 63

The setting of Fig. 26 (a) may be expressed simply as $\dfrac{C \mid b \mid c}{D \mid a \mid X}$ where the capitals C, D indicate as usual the slide rule scales on which the numbers a, b, c, X are read, and it is seen at once that the positions of the members of the proportion here are precisely the same as in the first of the two equations (28). In the same way we can set up each of the other three forms of the proportion expressions in (28) and (29).

Thus we have the general rule: "*Write any proportion* $a : b :: c : d$ *in the form* $\dfrac{a}{b} = \dfrac{c}{d}$ *and set up the two known members which occur together in the same relation on the C, D scales. The unknown is then found in line with the third known member.*" Thus the general form for pro-

C	a	c
D	b	d

Fig. 27

portion is Fig. 27 and the unknown member is found in the same position on the rule that it occupies in the proportion.

19. Squares, Cubes and Roots. — In connection with Figs. 10 and 13 we have already explained the meaning of the A scale and its relation to D. The B scale is a duplicate of A and its relation to C is the same as that of A to D. Thus squares and square roots may be found on B, C on the slide or A, D on the stock but A, D are generally more convenient and will here be referred to constantly.

In order to find the square of any number directly and by the use of the runner alone without the slide, simply set the runner hair line (indicator) on the number on D and at the indicator on A read its square. In order to find the square root of any number simply set the indicator to the number on A and at the indicator on D read its square root. These operations are so simple that they require no diagram or formal statement. The choice of scale (left or right) on A and the location of decimal points will require explanation but the settings and readings themselves offer no difficulties. Thus in Fig. 10 the two settings may be viewed as there indicated, namely, the finding of $\sqrt{9} = 3$ by reading from A to D and $6^2 = 36$ reading from D to A, or they may be viewed as the operations $3^2 = 9$, D to A, and $\sqrt{36} = 6$, A to D. Similarly in Fig. 13 the setting of the runner shows on A, D $\sqrt{17} = 4.13$ or $(4.13)^2 = 17$, and on B, C $\sqrt{12} = 3.46$ or $(3.46)^2 = 12$. Also, using the left C and B index line as an indicator we have on A, D $\sqrt{143.} = 1.195$ or $(1.195)^2 = 1.43$.

Concerning the decimal point and the choice of scale on A we may observe that the three settings already referred to may be read as we have written them above, namely:

$\sqrt{17} = 4.13$ $\sqrt{12} = 3.46$ $\sqrt{1.43} = 1.195$
$(4.13)^2 = 17$ $(3.46)^2 = 12$ $(1.195)^2 = 1.43$

or they may equally well have the following meaning:

$\sqrt{1700} = 41.3$ $\sqrt{1200} = 34.6$ $\sqrt{143} = 11.95$
$(41.3)^2 = 1700$ $(34.6)^2 = 1200$ $(11.95)^2 = 143$,

or even

$\sqrt{.17} = .413$ $\sqrt{.12} = .346$ $\sqrt{.0143} = .1195$
$(.413)^2 = .17$ $(.346)^2 = .12$ $(.1195)^2 = .0143$.

Thus whatever value we assign to the numbers on D, 1 to 10, 10 to 100, 100 to 1000, etc., or .1 to 1, .01 to .1, .001 to .01, etc., the numbers on A are their squares. Conversely, if the A numbers are properly assigned their values the D numbers are their square roots. It must be noted however that if the left index is taken as 1 the right index is 100 and not 10; the middle index is 10. Thus running from left to right the A scale will read 1 to 10 to 100, 100 to 1000 to 10,000, etc. and we cannot take the left A index as 10 for $\sqrt{10} = 3.16$ and 3.16 is not at the left index of D. In squaring numbers on D this question does not arise; we simply set the runner on the number on D and read its square on A. Thus numbers 1 to 3.16 on D have squares 1 to 10 on A, 3.16 to 10 on D have squares 10 to 100 on A. Coming back and starting again at the left index of D numbers 10 to 31.6 have squares 100 to 1000 on the left half of A, 31.6 to 100 on D have squares 1000 to 10,000 on the right half of A, etc. Thus when we know the position of the decimal point in a number on D the decimal point is automatically located in its square on A.

The same applies to decimal fractions. Reading the A, D scales right to left, if D runs 1 to .1, A runs 1 to .1 to .01; when D runs .1 to .01, A runs .01 to .001 to .0001, etc., and again the decimal point is automatically fixed in squares.

The proper half of A is selected for taking square roots by reversing the above considerations on squares of numbers. Thus for numbers between 1 and 10 use the left side of A and the root is between 1 and 3.16; between 10 and 100, right side of A and the root is between 3.16 and 10, etc. This method of selection does very well for fairly small numbers even though it requires a back-and-forth check or count of the scale indexes. It will of course also apply for very large numbers or very small decimal fractions but in such cases the counting and checking is very tedious. General rules can, however, be framed from these considerations which cover both the choice of A scale and the pointing off of squares and roots.

In extracting the square root of a number in arithmetic we begin at the decimal point and proceeding toward the left mark off the digits in pairs. Thus we write 144 as 1,44 and there is one digit before the decimal point in the root for each such group or part of a group in the number. As examples, $\sqrt{1,44.} = 12$. $\sqrt{56,25.} = 75$. $\sqrt{1,46,41.} = 121$. $\sqrt{81.} = 9$. With decimal fractions we do the same, proceeding toward the right. Thus $\sqrt{.01,46,41} = .121$ but $\sqrt{.14,64,10} = .383$ approximately, etc. We do the same when using the slide rule. Then after marking the groups toward the left in numbers greater than 1 we have this

> RULE: *If the last group contains one figure use A1 for finding the square root, if it contains two use A2. In either case the characteristic of the square root read on D equals the number of groups in the given number.*

As examples of the use of this rule consider the example discussed above. Thus $\sqrt{1,44.}$ is set on A1 and the root is 12. Similarly $\sqrt{29.8} = 5.45$ and the 29.8 is set on A2, and $\sqrt{12345.6}$ is set on A1 and the root is 111, approximately. For decimal fractions we obtain the following

> RULE: *After pairing off the digits to the right of the decimal point, at first disregard the groups immediately following the point which contain only ciphers. If in the first group containing other figures the first figure is a cipher use A1. If the first figure is not a cipher use A2. In the root there is one cipher immediately after the decimal point for each group consisting wholly of ciphers in the given number.*

Reversing the above discussion and rules for square roots we have for the decimal point in squares:

> RULE: *If the square is on A1 the characteristic of the square is 1 less than 2 times that of the number. if on*

$A2$ it is twice that of the number. This applies to both positive and negative characteristics.

A number may be squared more accurately though less conveniently by simply multiplying it by itself on the C, D scales. The decimal point is then located in the square by the use of the rule for multiplication. Similarly,

A	Read a^2	To a^2	Read a^3
B		Set 1	R to a
C			
D	Above a		

(a)

A		Read a^3
B		R to a
C	Set 1	
D	To a	

(b)

Fig. 28

reversing the operation of squaring, the square root may be found by setting the runner on the number on D and moving the slide until the same number is at the indicator on C and at the C index on D; this number is the square root of the original number.

The operation of finding the cube or cube root of a

number involves the B scale as well as A, D. To explain this write

$$a^3 = a^2 \times a.$$

With the number a set on D we can find a^2 on A as explained above. Using the B, A scales in the manner described in Article 14 we can then multiply a^2 by a and so find a^3. The complete operation is diagrammed in Fig. 28 (a).

It is to be noted, however, that when the B index is set at a^2 on A the C index is at a on D, so the C index may be set at a on D in the first place and the runner set at a on B without taking the intermediate reading a^2 on A. Thus the operation requires only one setting and with a set on D gives a^3 on A directly. The setting is given in Fig. 28 (b) and may be stated as follows: To find a^3,

(1) To a on D set C index,
(2) To a on B set runner,
(3) At indicator on A read a^3.

As an example $(4.1)^3 = 69$ is diagrammed in Fig. 29 (a). A second example is $(8.3)^3 = 571$, Fig. 29 (b).

If the left C index is set at 8.3 on D in the last example just given, it is found that 8.3 cannot be reached with the runner on either half of B. When the right C index is used, however, the runner may be set at 8.3 on either half of B and the reading on A is the same at both places, namely 571. In the first example, $(4.1)^3 = 69$, either

C index may be used. If the left index be used the left half of B is used with the right half of A and vice versa,

A		Read 69.0
B		R to 4.1
C	Set 1	
D	To 4.1	

(a)

A		Read 571.
B		R to 8.3
C	Set 1	
D	To 8.3	

(b)

Fig. 29

but the result is the same in both cases. Thus in different cases different combinations will be necessary but the diagram Fig. 28 (b) and the statement describing that setting will always hold good.

The decimal point is found in cubes of numbers by applying in succession the rules given above in this article for squares and in Article 16 for multiplication, or the following rule may be used:

>   RULE: *For numbers set on C to the left of 215 the characteristic of the cube is 2 less than three times that of the number; for numbers on C 215 to 465 it is*

SQUARES, CUBES AND ROOTS

1 less than three times that of the number; to the right of 465 on C, three times that of the number.

The finding of cube roots is the reverse operation of that of finding cubes. Thus if we take the setting for a^3 in Fig. 28 (b) as that of finding $\sqrt[3]{a^3} = a$ we notice that a^3 is set on A with the indicator and the slide is in such a

A	R to a	
B	Find $\sqrt[3]{a}$	
C	and	At 1
D		Find $\sqrt[3]{a}$

(a)

A	R to 91	
B	Find 4.5	
C	and	At 1
D		Find 4.5

(b)

Fig. 30

position that we have the same number at the indicator on B and at the C index on D. Thus we can diagram the cube root setting $\sqrt[3]{a}$ as in Fig. 30 (a) and state it as follows:

(1) Runner to a on A,
(2) Move slide until same number is at indicator on B and at C index on D,

72   THEORY AND OPERATION

(3) Runner to C index,
(4) At indicator on D read $\sqrt[3]{a}$.

The decimal point in cube roots is located by the following rule:

RULE: *Mark off the number whose cube root is to be found, in groups of three figures from the decimal point. The characteristic of the root equals the number of groups in the original number.*

A rule which applies to decimal fractions will be given later in connection with a cube scale.

The A and B scales may also be used in connection with C and D to find the squares and square roots of products and quotients and the products and quotients of squares and roots by other numbers. Consider the operation $(ab)^2 = X$. $(a \times b)$ is found as usual on D and immediately above this is its square on A. Similarly $(a \times b)$ may be found on A and $\sqrt{ab}$ is immediately below on D. Also $(a \div b)^2$ is on A above the quotient $(a \div b)$ which is found on D in the usual manner and $\sqrt{(a \div b)}$ is on D just below $(a \div b)$ on A. In all these it is not necessary to read the intermediate product or quotient but when the runner is set as if to read it on D or A the square or square root is read just above or below, as the case may be.

When the square or square root of any product, quotient or given number is found on A or D as explained above the product or quotient of this result by any other number may then also be found at once by simply moving the runner to the multiplier on B or C, or by setting the divisor on B or C to the runner in the position just found. If the operations for multiplication, division, squares and roots already given have been understood and practiced it is not necessary to give here settings or examples of these combined operations.

Since by formula (3) in Article 8, $a^4 = (a^2)^2$, the fourth power can be found by taking the square of the square. By (1) $a^5 = a^3 \times a^2$ and the fifth power is the product of the square and cube. Similarly $a^6 = (a^3)^2$ or $(a^2)^3$; $a^7 = a^4 \times a^3$; $a^8 = (a^4)^2$; $a^9 = (a^3)^3$; $a^{10} = (a^5)^2$; and so on.

Also by (9) $a^{2/3} = \sqrt[3]{a^2}$ and when a^2 is read on A from D then $\sqrt[3]{(a^2)} = a^{2/3}$ is found at once. Similarly $a^{3/2} = (\sqrt{a})^3$ and reading $\sqrt{a}$ on D and cubing the result $a^{3/2}$ is obtained at one setting.

19b. Use of Gauge Points on A and B Scales. — It has been pointed out that the gauge points marked on the A and B scales are at $\pi = 3.1416$ and $\frac{1}{4}\pi = 0.7854$. These are used in calculations involving the circumferences and areas of circles and in any other calculations involving π or $\frac{1}{4}\pi$ as factors or divisors.

We have already seen in Article 14 how A and B are used for multiplication. Thus to find $X = ab$ set B index to a on A and at b on B read X on A. Similarly the formula for the circumference c of a circle of diameter d is $c = \pi d$ and hence to find c set B index at π on A and at d on B read c on A. Similarly $d = \dfrac{c}{\pi}$ and to find the diameter when the circumference is known set π on B to c on A and at the B index read d on A. This calculation may also of course be carried out on C and D, and probably more accurately, but less conveniently.

The area of any circle of radius R is $a = \pi R^2$. Therefore to find the area when the radius is known set the C index to radius on D (which squares R on A) and at π on B read area on A. Similarly $a = \tfrac{1}{4}\pi d^2$ and to find the area when the diameter is known set C index to diameter on D and at the gauge point .7854 on B read the area on A. Both of these operations may of course be reversed to find $R = \sqrt{\dfrac{a}{\pi}}$ or $d = \sqrt{\dfrac{a}{.7854}}$ when the area is known.

In the same manner π or $\tfrac{1}{4}\pi$ may very conveniently be used as factors or divisors in calculations involving these numbers with any other numbers or squares.

20. The Inverted Slide. — If the slide is removed from the rule and replaced with ends reversed but with the same face (B, C scales) showing it is said to be "inverted." With the slide inverted the B scale is contiguous to D and C to A. The numbers on B, C have their usual significance and the usual decimal point rules

THE INVERTED SLIDE 75

apply. The use of the inverted slide simplifies certain operations and such uses will be briefly described. The same remarks apply to A and B as to D and C.

(i) *Reciprocals.* — Bring all indexes at each end into alignment with slide inverted. Any two numbers on C and D under the hair line of the indicator are now reciprocals, that is, their product equals 1 (or 10, 100, 1000, etc., with shifted decimal points). Thus to reduce any fraction of the form $\frac{1}{a}, \frac{10}{a}$, etc., to a decimal set the runner at the number on D (or C) and read the decimal under the indicator on C (or D).

(ii) *Multiplication and Division.* — Since in ordinary multiplication numbers on the C and D scales are in the same position and numbers read on C are the same as those on D, while now with the slide inverted C numbers are the reciprocals of those on D, it follows that if we set the index of the inverted slide on a D number and set the runner to a C number we are multiplying the D number by the reciprocal of the C number, that is, dividing by the C number. Thus the operations of multiplication and division are interchanged when the slide is inverted.

Suppose it is necessary to divide the same number by each of a series of different numbers. Ordinarily this requires a complete setting (slide and runner) for each division. With the inverted slide it is only necessary to set the slide index to the dividend on D and then set the runner on each of the divisors on C in turn and read the corresponding quotients on D. This follows from the fact that $D \div d = D \times \frac{1}{d}$.

The inverted slide also enables us to find all the pairs of corresponding factors of any given number. Thus the number $144 = 16 \times 9$, 18×8, 24×6, 30×4.8, 36×4, 48×3, 60×2.4, or 72×2, etc. This is shown in Fig. 31 with the slide inverted.

C Inv.	Set 1	Read 9	8	6	4.8	4	3	2.4	2
D	To 144	R to 16	18	24	30	36	48	60	72

Fig. 31

In order to find the factor pairs for any number, therefore, set the *Inverted C* index to the number on D. The runner at any factor on D then gives the corresponding factor on C.

(iii) *Inverse Proportion.* — Inverse proportion is a proportion in which two of the members are reciprocals of numbers instead of the numbers themselves. Thus if the direct proportion is

$$\frac{a}{b} = \frac{c}{d}, \text{ then } \frac{a}{b} = \frac{\left(\frac{1}{c}\right)}{\left(\frac{1}{d}\right)} \text{ or } \frac{a}{\left(\frac{1}{c}\right)} = \frac{b}{\left(\frac{1}{d}\right)} \quad (30)$$

is the inverse proportion. It may be briefly described by saying that while "in direct proportion more requires more and less requires less, in inverse proportion more requires less and less requires more." An illustration of a direct proportion is the problem: "If $2\frac{1}{2}$ tons of material costs \$35.50 what will $6\frac{3}{4}$ tons cost?" which is written $\frac{2.5}{35.5} = \frac{6.75}{X}$ and solved on the rule with slide in normal position as in Fig. 32 (a). A problem in inverse pro-

C Inv.	2.5	6.75
D	35.5	$x=95.75$

(a)

C Inv.	6	9
D	4	$x=2.67$

(b)

Fig. 32

portion is: "If 6 men can do a job in 4 days how long will it take 9 men to do it?" This is written in the form

$$\frac{6}{\left(\frac{1}{4}\right)} = \frac{9}{\left(\frac{1}{X}\right)}$$

and solved on the rule with inverted slide as in Fig. 32 (b). By comparing the formulas (30) and the settings Fig. 32 any inverse proportion is readily solved.

78   THEORY AND OPERATION

(iv) *Cube Roots.* — When the slide is inverted the positions of B and C are interchanged. The setting of Fig. 30 (b) therefore becomes Fig. 33 (a) and the statement of the operation for $\sqrt[3]{a}$ is:

   (1) To a on A set inverted C index,
   (2) Move runner until indicator stands on same number on B and D,
   (3) At indicator on D read $\sqrt[3]{a}$.

In Fig. 33 (b) is shown the setting for $\sqrt[3]{216} = 6$.

A	To a	
C	Set 1	
B		At $\sqrt[3]{a}$
D		Read $\sqrt[3]{a}$

(a)

A	To 216	
C	Set 1	
B		At 6
D		Read 6

(b)

Fig. 33

21. The Scale of Equal Parts (L) and Its Use. — The L scale is the scale of equal divisions which runs along the center of the reverse face of the slide. (The slide is said

THE SCALE OF EQUAL PARTS (L)   79

to be "reversed" when it is removed from the rule and replaced with faces interchanged but with indexes not interchanged.) When the slide is reversed and the indexes aligned the L scale runs from left to right and reads 0 to 1. The numbered divisions being tenths the larger subdivisions are hundredths and the smaller are two thousandths. Thus the L scale can be read directly to thousandths. A decimal point is always to be placed before the readings taken from this scale.

The L scale is shown at the center of the slide in Fig. 34 below. Comparing the L and D scales it is seen that they are related in the same manner as the rows of logarithms and numbers in the tables in Article 11. The numbers on L are therefore the logarithms of the numbers on D when the D indexes are 1 and 10, and the L indexes are in accordance with the formulas (23) in Article 9. When the D indexes are taken as 10 and 100, 100 and 1000, or 0.1 and 1, 0.01 and 0.1, etc., the figures on L are the mantissas of the logarithms of the numbers on D and the characteristics of the logarithms are found in the usual way (one less than the characteristic of the number, when the characteristic of the number is defined as above in Article 14).

Thus with the slide reversed and aligned we set the runner at any number on D and at the indicator read the mantissa of its logarithm on L, and vice versa. The L and D scales taken together therefore are the equivalent of a three-place table of logarithms. On this scale the base is 10, that is, the logarithms are common logarithms.

The L scale may also be used directly with the slide in the normal position by means of an indicator line at one end of the rule on the reverse side. In order to find logarithms in this manner draw the slide out toward the right until the number on C whose logarithm is desired is at the right D index. The logarithm (mantissa) is then on L at the indicator line on the reverse side of the rule and is read by simply turning the entire rule over. To find anti-logs set the mantissa on L at the indicator on the reverse side and read the anti-log on C at the right D index.

(NOTE. On some slide rules the L scale runs from right to left. To read logarithms or anti-logs on such a rule with reversed slide the slide must also be inverted. To use the slide in its normal position, when it is drawn out toward the right, the left C index is set at the number on D and the logarithm (mantissa) is read on the reverse indicator.)

22. The Sine Scale (S) and Its Use. — Besides the L scale there are also two other scales on the reverse face of the slide, the S and T scales. When the slide is reversed

THE SINE SCALE (S) AND ITS USE 81

and the indexes aligned the S scale lies along A and T along D. This arrangement is shown in Fig. 34.

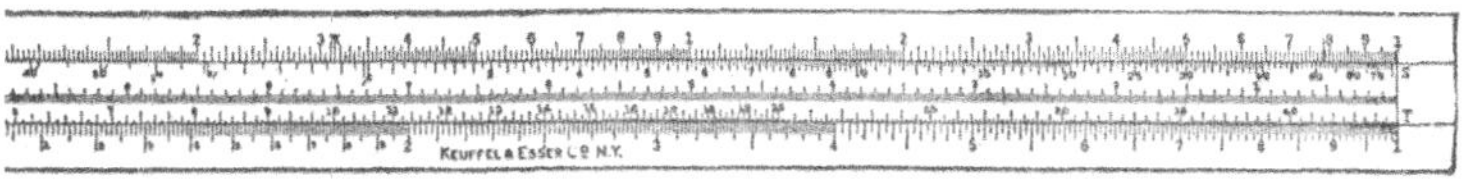

Fig. 34

The lengths laid off on S represent the logarithms of the sines of the angles 34 minutes to 90 degrees to the same plotting or drawing scale as that on which lengths on A represent the logarithms of the numbers 0.01 to 1. Thus with the indexes in alignment angles marked on S have their sines on A and vice versa. From 35' to 10° each of the smallest scale divisions on S represents 5'; from 10° to 20° each represents 10'; from 20° to 40°, 30'; 40° to 80°, 1°; and from 80° to 90°, 10°. Due to the fact that between 80° and 90° the sine changes very slowly as the angle changes the scale is greatly condensed in this range and the figures are not printed on the scale for 80° and 90°.

The characteristic of the natural sines read on A1 directly (angles 34.3' to 5° 45') is − 1; that is, the figures are written with a cipher between the decimal point and the first digit. The characteristic of the sines read directly on A2 is zero. Thus to find sin 2° 47' set the runner at 2° 47' on S and at the indicator on A1 read .0485. Similarly on A2 sin 13° 25' = .232.

For reading sines of angles less than 34.3' special gauge points are used. Just at the left of 2° on S is one of these marks, indicated by the minute sign ('); it is the "minutes" gauge point. Similarly the seconds gauge point (") is just at the right of 1° 10' line on S. To find the sine of any number of minutes set the minutes gauge point on S to the number of minutes on A and at the S index read the sine on A. To find the sine of any number of seconds set the seconds gauge point on S to the number of seconds on A and at the S index read the sine on A. These gauge points apply to both A1 and A2.

The characteristic of sines of angles 34.3' down to 3.43' is − 2; from 3.43' to .343' or 20.6" it is − 3; from 20.6" to 2.06" it is − 4; and from 2.06" to .206" it is − 5. As a convenient aid in locating the decimal point it should be remembered that sin 1' = .000292 (about 3 zeros 3) and that sin 1" = .00000485 (about 5 zeros 5).

The S scale may be used for multiplication in connection with A in the same manner as B is used. Thus to find $X = a \sin b$:

(1) To a on A set S index,
(2) To b on S set runner,
(3) At indicator on A read X.

This is diagrammed in Fig. 35 (a). In Fig. 35 (b) is set the multiplication $4.75 \sin 8° 40' = .715$.

A	To a	Read x
S	Set 1	R to b

(a)

A	To 4.75	Read .715
S	Set 1	R to $8° 40'$

(b)

Fig. 35

The decimal point is located in accordance with the usual rules, the characteristic of the sine factor being known from the statements given above. Division is also carried out in the usual manner, the dividends being numbers on A and the divisors being sines of angles on S.

The S scale may also be used to find sines and anti-sines with the slide in its normal position, by taking advantage of the indicator on the reverse side of the rule as described in Article 21 in connection with the L scale. Thus if the slide in normal position be drawn out toward the right, any angle at the reverse indicator on S has its sine at the right A index on B, and vice versa.

Since $\cos A = \sin(90° - A)$ the cosine of any angle is found by setting $(90° - \text{angle})$ on S and reading the sine on A by either method described above.

With the sine or cosine known we find the secant or cosecant from the relations

$$\sec = \frac{1}{\cos}, \quad \csc = \frac{1}{\sin}$$

by using the method for reciprocals, explained in Article 15. Thus with the slide in normal position set any angle A at the reverse indicator on S; the cosecant is then at the B index on A. Similarly set (90° − A) and the secant of A is at the B index on A.

The use of the S scale in the trigonometric solution of triangles will be explained in Chapter III

23. The Tangent Scale (T) and Its Use. — The lengths laid off on the T scale represent the logarithms of the tangents of the angles 5° 43′ to 45° to the same drawing scale as that on which the lengths on D represent the logarithms of the numbers .1 to 1. Thus with slide reversed and indexes aligned (Fig. 34) angles on T have their tangents on D, and vice versa. From 5° 45′ to 20° each of the smallest scale divisions on T represents 5′; from 20° to 45° each represents 10′. The characteristic of the natural tangents read on D directly is zero, that is, the figures read are immediately preceded by a decimal point.

Thus tangents are found by using T and D together in the same manner as S and A are used for sines. Multi-

THE TANGENT SCALE (T) AND ITS USE 85

plication and division of numbers by tangents of angles 5° 43' to 45° is carried out by using T and D in the same manner, including characteristics, as C and D are used in the multiplication and division of numbers, or as S and A are used with sines.

Tangents of angles smaller than 5° 43' are not given on T and D. Since to three figures, however, the tangent of an angle in this range is equal to its sine the S and A scales are used for tangents of these angles, and in exactly the same manner as for sines in every respect, including the characteristic.

The T scale may also be used in connection with the reverse indicator to find tangents and anti-tangents with the slide in its normal position. Thus if the slide in its normal position is drawn out toward the right, any angle at the reverse indicator on T has its tangent on C at the right D index, and vice versa.

Tangents of angles 45° to 90° are not given directly on T, D but since for any such angle A
$$\tan A = 1 \div \tan(90° - A),$$
that is, the reciprocal of $\tan(90° - A)$, it is only necessary to find $\tan(90° - A)$ on C with the slide in normal position as described above, and then to find the reciprocal of the number on D as explained in Article 15. Thus set at the indicator on the reverse side of the rule the angle $90° - A$. Then since $\tan(90° - A)$ is at the right D index on C,

tan $A = 1 \div \tan(90° - A)$ is at the left C index on D. This setting is therefore stated as follows:

(1) To reverse indicator set $(90° - A)$ on T,
(2) To C index set runner,
(3) At indicator on D read tan A.

It is diagrammed in Fig. 36 (a). Tangents of angles greater than 45° are greater than 1; the decimal point is located in accordance with the method of pointing off reciprocals. A numerical example, tan 52° 35' = 1.307, is diagrammed in Fig. 36 (b).

T (normal)	(90°−A) to Reverse Indicator	
C		R to 1
D		Read tan A

(a)

T (normal)	37° 25' to Reverse Indicator	
C		R to 1
D		Read 1.307

(b)

Fig. 36

Since $\cot = \dfrac{1}{\tan}$ the cotangent of any angle is found by a method similar to that described in Article 22 for the secant and cosecant. Thus for angles 5° 43' to 45° with the slide in normal position set the angle A on T at the

THE TANGENT SCALE (T) AND ITS USE   87

reverse indicator and at the D index on C read cot A. For angles 45° to 90° set 90° $- A$ on T at the reverse indicator and at the D index on C read cot A. For angles less than 5° 43′ use the S and A scales exactly as described for sines in Article 22.

The use of the T scale in the trigonometric solution of triangles will be described in Chapter III.

(NOTE. — Many useful conversion settings are stated and the settings for a number of useful and typical problems are diagrammed for the standard Mannheim slide rule in Chapter IV.)

CHAPTER III

MODIFIED FORMS OF THE MANNHEIM RULE AND THEIR USE

A. THE POLYPHASE MANNHEIM SLIDE RULE

24. The Polyphase Mannheim Slide Rule. — A brief description of the method of using the inverted slide has been given in Article 20. The conveniences of the inverted slide were recognized early in the development of the slide rule, the earliest record of its use being that of William Hunt in 1697. The regular logarithmic scale was soon afterwards used in the inverted position but was not printed on the rule in the inverted position (to read normally from right to left) until a hundred years later, in 1797. The form finally standardized by Mannheim and treated at length in Chapter II does not carry an inverted scale.

An inverted C scale has, however, been incorporated in the Mannheim rule by a number of designers, the resulting rules being in form the same as the standard Mannheim with the inverted C scale on the slide between the regular B and C scales. Several such rules are supplied in the United States under the names multiplex,

THE POLYPHASE MANNHEIM SLIDE RULE 89

maniphase, polyphase, etc., but all are essentially the same. One such form was produced by Keuffel & Esser about 1909 and is in very wide use.

Another scale which has been added to the Mannheim slide rule is a scale of cubes. It consists of three equal and similar parts laid out and divided in the same way as the A and B scales are formed. It is used in connection with the D scale for finding cubes and cube roots without setting the slide.

A Mannheim slide rule carrying both the inverted and cube scales is supplied by Keuffel & Esser and is called by them the "Polyphase (Mannheim) Slide Rule." This rule is illustrated in Fig. 37.

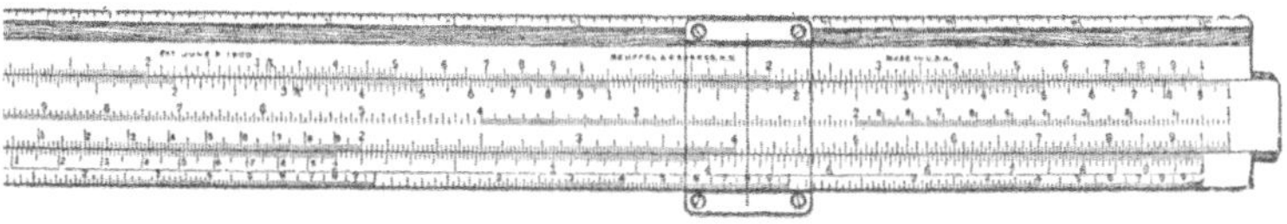

FIG. 37. POLYPHASE SLIDE RULE

The mechanical construction and form of the polyphase rule are the same as the standard Mannheim except that its width is about one eighth inch greater in order to provide space for the cube or K scale which is placed on the stock just below the D scale. (In some of the earlier forms of this rule the K scale is found on the edge of the stock instead of the face and the indicator is a line

90 MODIFIED FORMS OF THE MANNHEIM RULE

or mark placed on the frame of the runner in line with the hair line of the runner.) The slide is of the same width as the slide of the Mannheim rule and bears the same scales on its reverse face (S, L and T).

25. The Inverted C Scale and its Use. — The inverted C scale is placed on the front face of the slide between the B and C scales and is marked CI. It runs from right to left with the indexes aligned with those of B and C and the graduations are exactly the same as those of C. The figures on the CI scale are usually printed in red while those on the other scales are in black.

The principle involved in the use of the CI scale is discussed in Articles 15 and 20. In the actual use of the inverted slide as there explained, however, are several disadvantages. Thus the scale must be read upside down, the B scale cannot be used in its regular manner, and there is no normal C scale. In the polyphase rule the regular arrangement and its advantages are not disturbed but to these are added all the advantages of the inverted scale. The same rules for decimal points apply as with the regular C scale or the decimal point is located by inspection.

With the slide in its normal position the reciprocal of a number is read on CI (or C) by setting the runner at the number on C (or CI). Similarly decimal fractions are converted into common fractions with numerator 1 by setting the decimal fraction on C (or CI) and reading the denominator of the common fraction on CI (or C). Thus the reciprocal of 1.645 is $1 \div 1.645 = .608$. Similarly the decimal fraction .0555 is very nearly equal to the common fraction $\frac{1}{18}$ and the decimal fraction $.0532 = \frac{1}{18.2} = \frac{10}{182} = \frac{5}{91}$.

If the CI index is set on any number on D the product of every coinciding pair of numbers on CI and D, read at the indicator at the same setting, is equal to that number on D.

The operations of multiplication and division are interchanged when CI, D are used. Thus for *multiplication:*

(1) To *Multiplicand* on D set *Multiplier* on CI,
(2) To CI index set runner,
(3) At indicator on D read Product;

while for *division:*

(1) To *Dividend* on D set CI index,
(2) To *Divisor* on CI set runner,
(3) At indicator on D read Quotient.

Since the mechanical performance of these operations is familiar no examples or settings are given here.

In ordinary multiplication or division alone, however, the regular C, D scales are ordinarily used in the regular manner. The use of CI, D has the advantage, however, that when CI is set for the second factor the index is automatically set and it is not necessary to make a choice of index as is the case when using C, D.

The greatest advantage of the CI scale lies in the multiplication of three factors, which will readily be understood from the following considerations. The product

$$a \times b \times c = X \qquad (31)$$

may be written as

$$\left(a \div \frac{1}{b}\right) \times c = X. \qquad (32)$$

This is the same as the combined multiplication and division of formula (24), Article 17, and so is simply set on the slide rule according to Fig. 26 (a). The only difference in the operations corresponding to formulas (24) and (32) is that in (24) C is used to divide by b while in (32) CI is used to divide by $\frac{1}{b}$. Thus the operation corresponding to formula (31), to which (32) is equivalent, is stated as follows:

(1) To a on D set b on CI,
(2) To c on C set runner,
(3) At indicator on D read X.

This setting is diagrammed in Fig. 38 (a). It is much simpler than that shown and described for the same operation in Fig. 18 (c) in which only C, D are used.

THE INVERTED C SCALE AND ITS USE

An example of this operation is shown in Fig. 38 (b) which gives the setting for $25.7 \times 5.67 \times 4.12 = 600$.

CI	Set b	
C		R to c
D	To a	Read X

(a)

CI	Set 5.67	
C		R to 4.12
D	To 2.57	Read 600

(b)

FIG. 38

The method just described is readily extended to four or more factors. The formula

$$a \times b \times c \times d \times e = X \qquad (33)$$

can be written in the form

$$a \div \frac{1}{b} \times c \div \frac{1}{d} \times e = X$$

or as

$$\frac{a \times c \times e}{\left(\frac{1}{b}\right) \times \left(\frac{1}{d}\right)} = X$$

and in this form is seen to be of the same form as formula (25), Article 17. The operation corresponding to formula (33) is therefore carried out as described in Article 17 for formula (25) except that b and d are set on CI instead of C.

If a certain number is to be divided in turn by each of a series of different numbers (multiplied by their reciprocals) the method described in Article 20 for the inverted slide is used, viz., set the index to the constant dividend and then set the runner in turn on each of the divisors on CI, reading the corresponding quotients on D. Of course, if a single number is to be multiplied in turn by each of a series of different numbers the slide index is set to the number on D, the runner to the multipliers on C in turn, and the products read on D.

Although a very convenient method of finding the cubes of numbers will be described in connection with the scale of cubes a more accurate method is available with the CI scale used in connection with C and D. Thus the cube of any number a is $a^3 = a \times a \times a$, or $a \times a \times a = X$. This operation is exactly the multiplication of three numbers as shown in Fig. 38 (b) with b and c replaced by a. This may be stated as follows:

(1) To a on D set a on CI,
(2) To a on C set runner,
(3) At indicator on D read a^3.

Inverse proportion is handled with the CI scale in the same manner as with the inverted slide using CI for C Inv.

The settings described above cover the basic direct operations with the CI scale. Many other settings in which CI is used in connection with other scales are given in Chapter IV.

26. The Cube Scale and Its Use. — It has been shown in Articles 10, 11, 19 how the A and B scales give the squares of numbers on D and C respectively. The principle of these scales is explained in connection with Figs. 4, 5, 6 and it is there shown that the drawing scale to which A is laid off is one half of that for D, or in other words the logarithm laid off on a given length of A is twice the logarithm laid off on the same length on D. Similarly if another scale were to represent three times the logarithm on the same length on D it would be expected that the numbers on the new scale would be the cubes of the numbers on D. This is indeed the case and the K scale on the polyphase slide rule is such a cube scale. Conversely of course the numbers on D are the cube roots of the corresponding numbers on K.

As stated in Article 24 the K scale is adjacent to D on the face of the stock of the polyphase slide rule. It has four indexes, the first and fourth of which are aligned with those of D. The two intermediate indexes divide K into three equal and exactly similar parts or scales which will be referred to as K1, K2, K3, reading from left to right. Each part is divided and graduated exactly the same as the two parts of the A, B scales. Since each part is only two thirds the length of one half of A or B, however, readings cannot be made as closely on K as on A, B. When D is read as 1 to 10, then in order to give the cubes of D numbers K must be read as 1 to 1000 and the three parts as 1 to 10 to 100 to 1000. If D reads 10 to 100, K reads 1000 to 10,000 to 100,000 to 1,000,000; if D reads .1 to 1, K reads .001 to .01 to .1 to 1; etc. Thus when the decimal point is given in numbers on D the position of the decimal point in numbers on K is automatically fixed and at once known, and vice versa. Convenient rules will presently be given, however, which provide a ready means of pointing off cubes and cube roots of numbers as found on the K and D scales.

To find the cube of any number it is only necessary to set the runner at the number on D and its cube is at once read at the indicator on K.

To find the cube root of any number the runner is set at the number on K and its cube root is read at the indicator on D.

While there is only one place on D to set any particular number, however, there are three such places on K and these will give three different results for the cube root. Choice is made among these three by utilizing the method of cube root extraction of arithmetic in a manner analogous to that in which the square root method of arithmetic was used in Article 19 to determine the A scale settings for square roots. Thus we have the

RULE: *For numbers greater than 1 begin at the decimal point and mark off the number into groups of three figures proceeding toward the left. If the last group contains one, two or three figures use K1, K2, K3 respectively.*

If the number is a decimal fraction use the following

RULE: *Mark off the fraction in groups of three figures beginning at the decimal point and including ciphers. If the first group containing digits after the ciphers contains one, two or three such digits use K1, K2, K3 respectively.*

The decimal points in cubes and cube roots are located by the following rules which apply to positive or negative characteristics (numbers greater than 1 or decimal fractions):

RULE: *If the cube of a number is found on K1, K2, K3, to find the characteristic of the cube multiply that of the number by 3 and subtract 2, 1, 0 respectively.*

RULE: *If the cube root of a number is found from K1, K2, K3, to find the characteristic of the root add 2, 1, 0 respectively to that of the number and divide the result by 3.*

Due to the smallness of the divisions of the K scale readings cannot be taken so accurately as on A, B and C. The method for extracting cube roots described in connection with those scales in Article 19 is therefore some-

98   MODIFIED FORMS OF THE MANNHEIM RULE

times more useful. The most accurate method of all is that in which the full length C, CI and D scales are used, that is, the inverse operation of that described for cubes at the end of Article 25.

By using the K scale in connection with the other scales on the face of the rule any number may be raised to powers such as the following, either positive or negative:

$$\tfrac{2}{3}, \tfrac{3}{2}, 4, 5, 6, 9, \tfrac{5}{2}, \tfrac{5}{3}, \tfrac{5}{6}, \tfrac{4}{3}, \text{etc.}$$

and such expressions as the following may be evaluated at a single setting:

$$a \div \sqrt[3]{b^2},\ a \div \sqrt{b^3},\ a^2 \times \sqrt[3]{b^2},\ \sqrt{a^3 \times b^3} \div c^3, \text{etc.}$$

Some of these operations will now be explained and described.

By formula (9) in Article 8, $a^{2/3} = \sqrt[3]{a^2}$ or $(\sqrt[3]{a})^2$. If the number a is set on K then $\sqrt[3]{a}$ is on D and the square of $\sqrt[3]{a}$ is at the indicator on A. Therefore to find $a^{2/3}$ set the runner at a on K and read $a^{2/3}$ at the indicator on A.

Since by formula (5) in Article 8, $a^{-2/3} = 1 \div a^{2/3}$, when $a^{2/3}$ is found, then this result is set on C and $a^{-2/3}$ is read at the indicator on CI. Similarly $a^{3/2} = (\sqrt{a})^3$ and when a is set on A, $a^{3/2}$ is read on K and $a^{-3/2}$ can then be found with C and CI.

Since by formula (1) $a^5 = a^3 \times a^2$ the fifth power is found at one setting in the following manner which is illustrative of the possible methods of combination of settings:

DUPLEX RULE AND THE FOLDED SCALES 99

(1) To a on D set runner and C index,
(2) At indicator on K read a^3 and set runner to this on B,
(3) At indicator on A read a^5.

Thus a^2 on A is multiplied by a^3 on B as read from K. It is not necessary to read a^2 on A since when the C index is set at a on D the B index is thereby set at a^2.

By formula (3) $a^6 = (a^3)^2 = (a^2)^3$, hence the sixth power is found as the square of the cube or the cube of the square. Similarly the fourth power is the square of the square and the ninth is the cube of the cube.

Other positive fractional powers are found by applying formula (9) with the K, A, D scales and the negative powers are the reciprocals of the positive powers and are found with C and CI. Thus

$$a^{5/2} = \sqrt{a^5}, \qquad a^{5/3} = \sqrt[3]{a^5},$$
$$a^{5/4} = \sqrt{a^{5/2}}, \qquad a^{5/6} = \sqrt[3]{a^{5/2}}, \text{ etc.}$$

The remaining expressions mentioned above together with many others are evaluated by means of settings stated and diagrammed in Chapter IV.

B. THE POLYPHASE DUPLEX SLIDE RULE

27. The Duplex Rule and the Folded Scales. — Although the slide of the Mannheim slide rule has scales graduated on both faces the stock is graduated on only one face. Slide rules having scales on both faces of the stock as well as the slide are called "duplex" rules. We

100 MODIFIED FORMS OF THE MANNHEIM RULE

have seen that the first duplex slide rule was that produced by Seth Partridge in 1657. The form of duplex now in common use in the United States was brought out by William Cox in 1891 and uses the standard Mannheim scales. The slide moves between two outer strips of wood which are held together by thin metal cleats and all three strips are of the same thickness, so that their faces are flush and scales may be graduated on both faces of all three strips.

In 1817, as we have seen, an arrangement called the "folded scale" was introduced. In this arrangement the scale is "folded" near the center and the indexes placed together. The scale is then parted at the folding point and the separated ends are made the ends of the new scale, while the now coincident original indexes form a single index near the center of the new scale. This arrangement has been applied to the Mannheim scales by French manufacturers and by Keuffel & Esser in the United States. A duplex rule with the folded scales and all the regular scales of the polyphase Mannheim rule is supplied by Keuffel & Esser and is known as the "polyphase duplex" slide rule. This form of rule is widely used among engineers in the United States.

The form and scales of the polyphase duplex slide rule are shown in Fig. 39, in which (a) shows the main face or front of the rule and (b) the reverse face or back of the rule.

DUPLEX RULE AND THE FOLDED SCALES 101

As seen in Fig. 39 (a), the front of the stock carries the regular D scale and the folded D scale, called DF; the front of the slide carries the regular and folded C scales, C and CF, and the folded CI scale, CIF. The back of the stock carries the regular L, D, A and K scales and the back of the slide the regular CI, T, S and B scales.

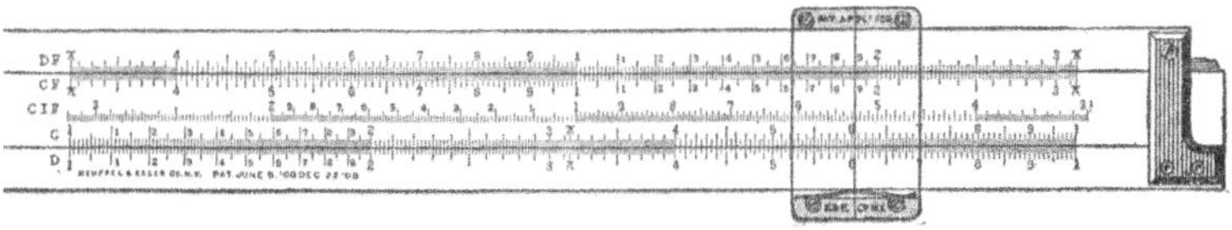

(a) Front

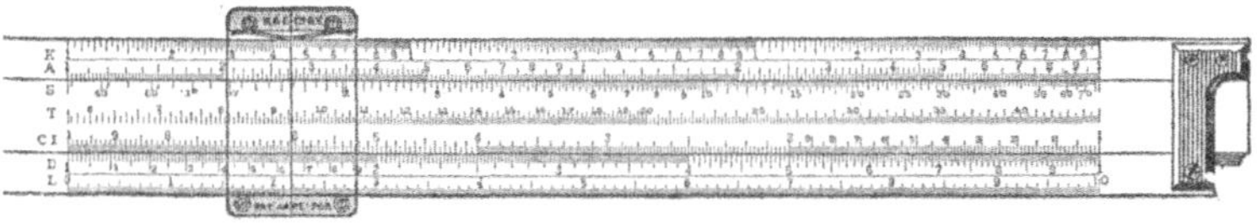

(b) Back

FIG. 39. POLYPHASE DUPLEX SLIDE RULE

The D, A and B scales are in their usual positions but L is now fixed adjacent to D with which it is used and CI is also adjacent to D with which it is used. The S scale is also fixed adjacent to its associated scale B or may be set in connection with A, and the T scale is conveniently placed with regard to D, with which it is used. On the front C and D are in their usual positions and immediately above are their associated scales CF and DF, D and DF being in a fixed relation on the stock while C and CF are fixed on the slide. Similarly CF and its invert CIF are in a fixed relation on the slide, as are C and CI on the opposite face.

The runner extends around the entire rule and carries two indicator lines which are exactly aligned. With this arrangement not only can any two scales on the same face be read together, but also any two scales on opposite faces; thus any or all of the scales may be read together. This arrangement reduces the number of settings required in complicated operations and facilitates the use of the rule in general.

Instead of being folded at the center which is $\sqrt{10} = 3.1623$, the C, CI, D scales are folded at $\pi = 3.1416$ to give CF, CIF and DF. Thus the C and D indexes are aligned with π on CF and DF respectively and the middle indexes of CF, DF with π on C, D respectively. Now we have already seen that with the C index set at any number on D any two coinciding numbers on C, D are in the ratio equal to the number at which the C index is set. Therefore the ratio of all aligned numbers on DF and D and on CF and C is equal to π, and any number on DF or CF is π times its corresponding number on D or C respectively.

28. Operation of the Polyphase Duplex Slide Rule. — The A, B, C, D and L, S, T scales of the polyphase duplex rule are the regular Mannheim scales and their use is precisely the same as described in Chapter II. The K and CI scales are the same as on the polyphase (Mannheim) rule and their use is described above in Articles 25 and 26. The use of these scales need not therefore be described here and will be referred to only in connection with the folded scales and the new combinations made possible by the rearrangement of their positions.

In tne description of the polyphase duplex rule in the last Article (27) the multiplication of numbers by π is explained, viz., set the runner at the number on C or D and read π times the number on CF or DF respectively at the indicator. In the same way any number on CI is aligned with π times the number on CIF. These settings are illustrated in Fig. 39. Thus in (a) the indicator is set at 6 on D and at the indicator on DF is $6\pi = 18.85$. Also (since the slide and stock indexes are aligned in Fig. 39) 6 on C is aligned with $6\pi = 18.85$ on CF. Conversely there is found on C or D the quotient by π of any number on CF or DF respectively. Thus we may consider the indicator in Fig. 39 (a) as set at 18.85 on DF and on D find $18.85 \div \pi = 6$, and similarly on CF and C.

These scales furnish a very convenient table of diameters (d) and circumferences (c) of circles, for $c = \pi d$ and $d = \dfrac{c}{\pi}.$ If the runner is set at any diameter on D the corresponding circumference is read at the indicator on

DF, and if it is set at any circumference on DF the corresponding diameter is read on D. C and CF may of course be used in the same way.

Any number which can be read on C or D in connection with any of the regular scales (products, quotients, cube roots, square roots, etc.) can be immediately multiplied by π by the use of the folded scales. Thus if any number a is set on K, then $\pi\sqrt[3]{a}$ is read at the indicator on DF directly, and if a is set on A then $\pi\sqrt{a}$ is read directly on DF. If $a \times b = x$ is set in the usual manner on C, D, then since x is on D, πx is at the same position of the indicator on DF. Since $\pi x = \pi ab$ we have the following setting for $\pi ab = X$:

(1) To a on D set C index,
(2) To b on C set runner,
(3) At indicator on DF read X.

Similarly for the operation $\dfrac{\pi a}{b} = X$:

(1) To a on D set b on C,
(2) To C index set Runner,
(3) At indicator on DF read X.

Similarly also such operations as

$$\frac{\pi ac}{b}, \frac{\pi ac}{bd}, \frac{\pi ace}{bd}, \text{etc.,}$$

can be performed in the same general manner. Simply make the settings on C, D, CI in the usual manner as for

$$\frac{ac}{b}, \frac{ac}{bd}, \frac{ace}{bd}, \text{etc.,}$$

OPERATION OF THE POLYPHASE DUPLEX RULE 105

and when the runner is brought to its final position read on DF instead of on D. This means that the result of the C, CI, D setting is multiplied by π.

Since CF and DF are exactly alike these two scales may be used for multiplication and division in the same manner as C and D or A and B are used together. It is only necessary to remember that the indexes of CF and of DF lie together near the centers of the scales instead of at the ends. Thus to carry out the multiplication $3 \times 6 = 18$, at 3 on DF set the (middle) CF index and at 6 on CF read the product 18 on DF. To perform the division $8.4 \div 4 = 2.1$, at 8.4 on DF set 4 on CF and at the (middle) CF index read the quotient 2.1 on DF.

The scale CIF is the reciprocal or invert of CF. Therefore in precisely the same manner as on C, D and CI the operations

$$\frac{a}{b}, \frac{ac}{b}, \frac{ac}{bd}, \frac{ace}{bd}, \text{etc.},$$

can be carried out on CF, DF and CIF and then, since any result on DF is divided by π on D, by making these settings on CF and DF and reading on D at the final position of the runner instead of on DF the results

$$\frac{a}{\pi b}, \frac{ac}{\pi b}, \frac{ac}{\pi bd}, \frac{ace}{\pi bd}, \text{etc.},$$

are obtained.

It can be said in general, therefore, that any operation

involving π as a factor can be carried out on scales C, D in the usual manner without π and the result read on DF instead of D at the final position of the runner; and vice versa any operation involving π as a divisor can be carried out on CF and DF in the usual manner without π and the final result read on D instead of DF.

A useful particular case of this general statement is that of reciprocals. Reciprocals of numbers are found on the polyphase duplex rule as on the polyphase Mannheim but the C and CI scales are on opposite faces of the rule and when a number is set on C the rule is turned over to read the reciprocal on CI. This operation can be performed even more simply on CF and CIF since these scales are reciprocal and are adjacent on the front of the rule. Also, since if the number a is set on C then πa is read on CF and the reciprocal of this quantity on CIF, the expression $\frac{1}{\pi a}$ is read directly on CIF when a is set on C. Similarly if a is set on CF $\frac{\pi}{a}$ is read on CI. For, when a is on CF then $\frac{a}{\pi}$ is on C and $1 \div \left(\frac{a}{\pi}\right) = \frac{\pi}{a}$ is on CI.

Another very useful application of the folded scales (in reality the purpose for which the folded scales were designed) is illustrated by the following example.

$$\frac{52 \times 7}{13} = 28$$

OPERATION OF THE POLYPHASE DUPLEX RULE 107

The usual procedure is

(i)
(1) To 52 on D set 13 on C,
(2) To 7 on C set runner,
(3) At indicator on D read 28.

When the attempt is made to carry out step (3), however, it is found that the setting is "off the scale" and the usual procedure is to exchange C indexes and then go ahead with steps (2) and (3); so that the operation (i) really amounts to

(ii)
(1) To 52 on D set 13 on C,
(2) To C index set runner,
(3) To runner set other C index,
(4) To 7 on C set runner,
(5) At indicator on D read 28.

In this form the operation consists of five steps and is both inconvenient and liable to error. By the use of the folded scales, however, form (ii) can usually be avoided and form (i) used as if step (2) did not "run off the scale." Thus

(iii)
(1) To 52 on D set 13 on C,
(2) To 7 on CF set runner,
(3) At indicator on DF read 28.

For this form of the setting to be used it is generally (though not always) necessary for at least half of the length of the slide to be inside the stock. Thus the operation $\frac{52 \times 9}{13} = 36$ could not be carried out by either of the methods (i) or (iii) but by using CF and DF first and then C and D it becomes simply

(iv)
(1) To 52 on DF set 13 on CF,
(2) To 9 on C set runner,
(3) At indicator on D read 36.

With practice any setting such as (i) or (ii) and their extensions can be replaced by one like (iii) or (iv) and much time saved. In general, in any operation in which a setting of the runner on C "runs off the scale" set it on CF instead and read on DF, and vice versa. It is easily seen that this is possible because the folded scales are graduated exactly like the regular scales but are so arranged that, in effect, the portion of the regular scale which projects beyond the reach of the runner is doubled back so as to be inside the stock.

The direct uses of the folded scales in connection with the standard (polyphase) scales have now been described. In the same way as for combination settings of the standard scales, so also combination settings of these with the folded scales can be multiplied indefinitely. A large number of such settings are given for the polyphase duplex slide rule in the next chapter.

C. THE LOG-LOG DUPLEX SLIDE RULE

In order properly to understand the so-called "log-log" slide rule it is necessary to extend the short discussion of exponents and logarithms given in Articles 8 and 9.

NATURAL LOGARITHMS AND EXPONENTIALS 109

In general the common logarithms, or logarithms to base 10 are used and the standard slide rule scales (including inverted and folded scales) are based on those logarithms. In higher mathematics, however, and in steam, gas, hydraulic and electrical engineering the so-called "natural" logarithms (to base 2.7183) are used, together with expressions involving powers of this base called "exponentials." We will here state briefly some of the properties of natural logarithms and exponentials and their relations to common logarithms. As in Articles 8 and 9 the discussion will be more descriptive and explanatory than demonstrative and the reader who cares to study these subjects is referred to any good textbook of algebra or analytical trigonometry.

29. Natural Logarithms and Exponentials. — As implied in the definition of a logarithm in Article 9 logarithms may be referred to any number as base, the base 10 being used in ordinary calculation because of its convenience. In the theoretical investigations of higher mathematics and in the formulas of certain branches of physics and engineering it is more convenient to use the base mentioned above, which is designated by the letter "e." (In certain work in electricity where "e" is used to indicate electromotive force the Greek letter "epsilon" (ϵ) is used.) This number is in reality, like π, an unending decimal

($e = 2.718281828 + \ldots$) which makes its appearance in certain transformations in higher mathematics as the definite value to which the sum of the unending series

$$1 + \frac{1}{1} + \frac{1}{1 \cdot 2} + \frac{1}{1 \cdot 2 \cdot 3} + \frac{1}{1 \cdot 2 \cdot 3 \cdot 4} + \ldots$$

becomes more nearly equal as the number of terms is continually increased. Logarithms referred to the base e are called "natural logarithms." Conversion of logarithms from the base e to the base 10 and 10 to e, or from any base to any other are readily made.

Thus let A be any number and let a and b be the bases of any two systems of logarithms. Then by the definition of a logarithm and by formula (14) of Article 9 we can write

$$m = \log_b A, \qquad n = \log_a A, \qquad (34)$$
$$b^m = A, \qquad a^n = A,$$
$$a^n = b^m.$$

Taking the logarithm of both sides of this last equation to the base a we get, by formula (17),

$$n \log_a a = m \log_a b.$$

But by (23) $\log_a a = 1$, therefore this becomes

$$n = (\log_a b) \cdot m.$$

Putting in this the values of m and n from (34) gives finally

$$\log_a A = (\log_a b) \cdot \log_b A. \qquad (35)$$

Hence if the logarithm of A to the base b is known and the logarithm of the same number A to some other base a

NATURAL LOGARITHMS AND EXPONENTIALS 111

is desired it is only necessary to multiply the known logarithm by the logarithm of b to the base a. From (35) we have also

$$\log_b A = \left(\frac{1}{\log_a b}\right) \cdot \log_a A. \qquad (36)$$

Thus suppose $\log_e A$ appears in a formula and no tables of logarithms to base e are available. By means of (36) $\log_{10} A$ may be used instead. In this case $a = 10$, $b = e = 2.7183$ and the formula (36) is

$$\log_e A = \left(\frac{1}{\log_{10} 2.7183}\right) \cdot \log_{10} A.$$

From a table of common logarithms it is found that the quantity $\log_{10} 2.7183 = 0.43429$; also $\frac{1}{.43429} = 2.3026$, and the conversion formula finally becomes

$$\log_e A = 2.3026 \log_{10} A. \qquad (37)$$

Similarly (35) becomes

$$\log_{10} A = .43429 \log_e A. \qquad (38)$$

For purposes of slide rule transformation (with logs to base 10 found on the L scale) the multiplier in (37) may be taken as 2.3; similarly the multiplier in (38) is easily remembered as .434.

By means of (37) 2.3 times the common logarithm is used in any formula for the natural logarithm; in this conversion both the characteristic and the mantissa of the common logarithm are multiplied by 2.3 to give the natural logarithm. It is obvious that the characteristic of the natural logarithm thus found bears no simple relation to the number of digits in the given number A and is therefore not readily determined by inspection. For this reason tables of natural logarithms contain both mantissa and characteristic, the complete logarithm of every number in the table, with the decimal point properly placed. Such tables are given in handbooks of chemistry, physics and engineering.

The formulas of Article 9 apply to logarithms of any base (10, e, Napier's base, etc.). Thus from (23)
$$\log_e 1 = 0, \quad \log_e e = 1. \tag{39}$$
If the symbol ∞ is defined as a quantity infinitely great ("infinity") then we may write as limiting values
$$\frac{1}{\infty} = 0, \quad b^\infty = \infty$$
and hence
$$\frac{1}{b^\infty} = 0, \quad \text{or} \quad b^{-\infty} = 0.$$
If we let b represent any base and take the logarithm of both sides of the last equation to the base b, then by formula (14)
$$\log_b 0 = -\infty$$
If $b = e$,
$$\log_e 0 = -\infty. \tag{40}$$

From the definitions of negative exponents and of logarithms it is obvious that the common logarithms of

NATURAL LOGARITHMS AND EXPONENTIALS   113

decimal fractions (numbers between 0 and 1) are negative; it can be shown that this is the case with any base, in the same manner. This is illustrated by (40) and the first of formulas (39). Thus we have that all numbers above 1 have positive logarithms, the logarithm of 1 itself is zero, and in passing from 1 to zero (through all fractions) the logarithm goes from zero to minus infinity, that is, includes all negative numbers. Since the positive numbers 0 to $+\infty$ thus have all logarithms $-\infty$ to $+\infty$, negative numbers have no logarithms in the sense in which they have been defined here. Using the symbol $<$ to mean "less than" (and similarly $>$ would mean "greater than") then it can be said in general that if

$$0 < A < 1, \quad \log_e A < 0, \qquad (41)$$

that is, negative. Of course this applies to any other base as well as to e.

If in the equation $3x + 2 = 8$, x is unknown the equation is called a "first degree" equation; $2x^2 - 7x + 5 = 0$ is a "second degree" equation. In each of these the unknown quantity is a base and the equation is described by reference to the exponent. Similarly such an equation as $4^x = 64$ in which the unknown quantity is the exponent is called an "exponential" equation and the quantity 4^x is the exponential. In particular are such expressions as e^x, e^{-x}, e^2, etc., always known as "ex-

114 MODIFIED FORMS OF THE MANNHEIM RULE

ponentials." Exponentials occupy a large place in the mathematics of physics and engineering.

The exponential equation
$$4^x = 64$$
is solved as follows: Take the logarithm (to any base) of both sides by (17),
$$x \log 4 = \log 64$$
$$\therefore x = \frac{(\log 64)}{(\log 4)}.$$

If the logarithms are common logarithms (as is the case in ordinary arithmetical calculations), then this expression gives
$$x = \frac{1.806}{.602} = 3$$
when the logarithm mantissas are read to three figures on the L scale of the slide rule. Of course it may be seen at once without using logarithms that when 4 to a certain power equals 64 then the power must be the cube, but in a case like
$$4.76^x = 2.19 \tag{42}$$
the value of x is not immediately obvious. Exactly as in the previous case, however,
$$x = \frac{(\log 2.19)}{(\log 4.76)} = \frac{.340}{.678} = .502; \tag{42a}$$
and in general if
$$b^x = a, \quad x = \frac{\log a}{\log b}. \tag{43}$$

EXPONENTIALS, POWERS AND ROOTS 115

As in the case of the base b so with the base e in equations (1) and (2) in Article 8,

$$e^x \times e^y = e^{x+y} \tag{44}$$

$$e^x \div e^y = e^{x-y}. \tag{45}$$

30. Exponentials, Powers and Roots, and the Slide Rule. — In order to carry out the solution of the exponential equation (42) as in (42a) by means of the slide rule the procedure is clearly as follows:

$$\text{I} \begin{cases} \text{(i)} & \text{Find } \log 2.19 = .340; \\ \text{(ii)} & \text{Find } \log 4.76 = .678; \\ \text{(iii)} & \text{Divide } \dfrac{.340}{.678} = .502. \end{cases}$$

These three operations (on Mannheim, polyphase or duplex rule) are:

(i) $\begin{cases} (1) \text{ To 219 on D set runner,} \\ (2) \text{ At indicator on L read 340,} \\ (3) \text{ Charact. of log is 0, write .340;} \end{cases}$

(ii) $\begin{cases} (1) \text{ To 476 on D set runner,} \\ (2) \text{ At indicator on L read 678,} \\ (3) \text{ Charact. of log is 0, write .678;} \end{cases}$

(iii) $\begin{cases} (1) \text{ To .340 on D set .678 on C,} \\ (2) \text{ To C index set runner,} \\ (3) \text{ At indicator on D read .502.} \end{cases}$

In the same way the solution of (43) can be carried out for any values of a and b.

Since in (43) a, b and x may have any values consistent with one another we can also write the similar equation

$$A^B = x \tag{46}$$

with x unknown and can also solve this equation logarithmically. Thus as an illustration let us consider
$$13.65^{2.37} = x. \qquad (46a)$$
As before $2.37(\log 13.65) = \log x$. To find x then we must

II $\begin{cases} \text{(i)} & \text{Find } \log 13.65 = 1.135; \\ \text{(ii)} & \text{Multiply } 2.37 \times 1.135 = 2.69; \\ \text{(iii)} & \text{Find antilog } 2.69 = 186.0 \end{cases}$

Each of these three operations is easily carried out on the C, D and L scales of the slide rule.

Consider next an equation of the form
$$x^a = B, \qquad (47)$$
for example,
$$x^{3.62} = 79.5; \qquad (47a)$$
As before
$$3.62(\log x) = \log 79.5;$$
$$\therefore \log x = \frac{(\log 79.5)}{3.62},$$
and the slide rule solution is

III $\begin{cases} \text{(i)} & \text{Find } \log 79.5 = 1.900; \\ \text{(ii)} & \text{Divide } \dfrac{1.900}{3.62} = .525; \\ \text{(iii)} & \text{Find antilog } .525 = 3.35. \end{cases}$

The three operations (i), (ii), (iii) of each of the complete procedures II and III are separately simple and similar to those of I as described above and will not be given here in detail.

If the equation (47a) be compared with equations (18) and (7) it will be found that in solving (47a) we have taken the 3.62th root of 79.5, that is, $x = \sqrt[3.62]{79.5} = 3.35$. Similarly in (46a) we have raised 13.65 to the 2.37th power, that is, $x = 13.65^{2.37} = 186.0$. In general, therefore, the slide rule can be used to find any power or root of any number by the ordinary logarithmic method, but as has appeared the operations require many settings. By means of the so-called "log-log" scale (or scales) used in connection with the ordinary Mannheim scales all three operations (i), (ii), (iii) of either procedure I, II or III can be performed at one setting.

Since equations (46) and (47) are only modifications of (43) a scale designed to solve (43) can be used to solve (46) and (47) in the same general way that C and D designed originally for multiplication can also be used for division and reciprocals. The log-log scale will therefore be described on the basis of equation (43) and then shown to apply to (46) and (47). It will also be seen that it furnishes a table of natural logarithms and exponentials.

31. The Log-Log Scale and the Log-Log Duplex Slide Rule. — We re-write here equations (43),

$$b^x = a, \qquad x = \frac{(\log a)}{(\log b)}, \tag{43}$$

and for comparison also write equation (16) in the form

$$x = \frac{A}{B}, \quad \log x = \log A - \log B. \tag{16}$$

In order to find $x = \dfrac{A}{B}$ by means of the slide rule we have found that according to (16) two scales of logarithms *of numbers* are necessary and the difference between the lengths corresponding to $\log A$ and $\log B$ is the length corresponding to $\log x$, and the numbers being printed on the scales at the points corresponding to the scale lengths of their logarithms, A and B can be set on the scales and x read directly.

The scales necessary for the solution of (43) are now clearly defined: they must be laid off in lengths corresponding to logarithms *of logarithms*. Thus from (43)

$$\log x = \log(\log a) - \log(\log b) \qquad (43a)$$

and if on a scale marked with logarithms of logarithms we find the difference between $\log(\log a)$ and $\log(\log b)$ and then lay off this distance on the ordinary scale of logarithms of numbers, this last distance will represent $\log x$ and the scale number will be x. The new scale of logarithms *of logarithms* is the "log-log" scale and the ordinary logarithmic scale of numbers is the regular Mannheim C scale. Something of the history of the origin and adoption of the log-log scale is given in Articles 4 and 7.

In construction the log-log scale is laid off in three or four sections on the stock of a duplex slide rule. This requires that the wooden strips forming the stock be

THE LOG-LOG SCALE

somewhat wider than in the polyphase duplex; otherwise the form is the same as that of the polyphase duplex rule. A complete rule formed in this manner is supplied by Keuffel & Esser and is called by them the "Log-log Duplex" slide rule. It is the same as the polyphase duplex rule of these manufacturers with the log-log scales added, as appears in Fig. 40, in which (a) shows the front and (b) the back of the complete rule.

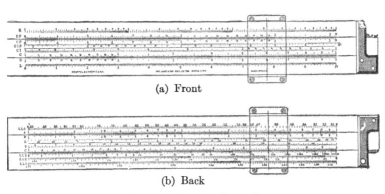

(a) Front

(b) Back

FIG. 40. LOG-LOG DUPLEX SLIDE RULE

As seen in Fig. 40 (a) the front of the stock of the log-log duplex slide rule carries regular and folded scales L, D, DF, K, and the front of the slide the C, CI, CIF, CF scales. In Fig. 40 (b) are seen on the back of the slide the standard B, S, T, C scales and on the back of the stock, the standard A scale and the log-log scale. marked LL and divided into four sections, LL0, LL1, LL2, LL3.

LL0 is above A and B in connection with which it is used, and LL1, LL2, LL3 are below C, in connection with which they are used, and their indexes are aligned with the right and left indexes of the stock.

The three LL sections, 1, 2, 3, if placed end to end would form a continuous scale from $e^{.01} = 1.01005$ to $e^{10} = 22026.3$. LL1, placed at the lower edge of the stock, reads from left to right $e^{.01} = 1.01005$ to $e^{.1} = 1.10517$; LL2 is just above LL1 and reads from left to right $e^{.1} = 1.10517$ to $e = 2.71828$; and LL3, contiguous to C, reads from left to right $e = 2.71828$ to $e^{10} = 22026.3$. Exponentials greater than e^{10} are rarely used so a section LL4 is not added. The right index of LL2 and the left index of LL3 are marked "e."

As shown in Article 29 above, equations (39), (40), (41), $\log_e 1 = 0$ and therefore $\log_{10}(\log_e 1) = -\infty$, and since $\log_{10}(\log_e 1.01) = -4.61015$, the difference between the log-log of 1.01 and of 1.00 is infinite and it would require a scale of infinite length to cover this range. LL1 therefore does not extend below log-log 1.01.

LL0 covers the range from $e^{-.03} = \dfrac{1}{e^{.03}} = .970$ to $e^{-3} = \dfrac{1}{e^3} = .050$ and is so arranged that e^{-1} is aligned with the right and left indexes of the A scale and $e^{-1} = \dfrac{1}{e} = .368$ is aligned with the middle A index.

USE OF THE LOG-LOG SCALES   121

The decimal point is already placed in all the numbers on LL and the scale divisions are decimal and self-explanatory. The use of all scales on the log-log duplex slide rule other than LL has already been described in connection with the other rules and will not be discussed here except as they are used in connection with LL.

32. Use of the Log-Log Scales. — In order to explain the use of the log-log scales equations (43) and (43a) are here re-written for convenient reference:

$$b^x = a, \quad \log x = \log(\log a) - \log(\log b). \quad (48)$$

The method of finding x from the last of these equations by means of the logarithmic scales is shown in Fig. 41.

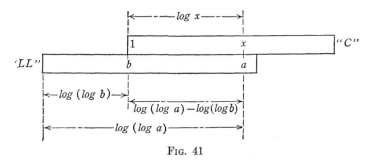

Fig. 41

On LL at a distance from the left index equal to $\log(\log b)$ is marked the number b, and at a distance from the left index equal to $\log(\log a)$ is marked the number a. The difference between these two lengths is $\log(\log a)$

$- \log(\log b)$. But according to equation (48) above, this is equal to $\log x$. Now the C scale is laid off in lengths equal to the logarithms of the numbers printed on it and at a distance from the left index on C equal to $\log x$ is marked the number x. If therefore the left index of C is set at b on LL as shown in Fig. 41 then x on C will be at a on LL. Thus to solve the exponential equation in (48), set the left C index at b on LL and at a on LL read x on C. As an example let us solve the simple equation already discussed:

$$4^x = 64.$$

The setting for this operation is stated as follows:

(1) To 4 on LL set C index,
(2) To 64 on LL set runner,
(3) At indicator on C read 3.

This setting is diagrammed in Fig. 42.

C	Set 1	Read $x=3$
LL3	To 4	R to 64

Fig. 42

As has already been found in Article 30 the equation (46),

$$A^B = x, \qquad (46)$$

is logarithmically of the same form as the equation $b^x = a$; the only difference lies in the fact that the unknown quantity x occupies a different position. Therefore

(46) is solved by the same type of setting as that just explained with the steps taken in a different order. It will here be explained independently, however. Applying the logarithmic formula (17) of Article 9 to (46) above we get
$$B(\log A) = (\log x)$$
and on applying to this the logarithmic formula for a product
$$\log B + \log(\log A) = \log(\log x) \qquad (49)$$
or
$$\log B = \log(\log x) - \log(\log A).$$

This form shows that (46) is equivalent to the equation already solved and explained. For separate solution, however, it is used in the form (49). The setting for this solution is given in Fig. 43.

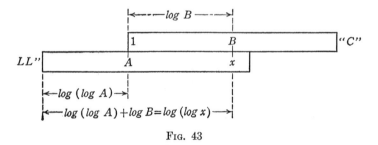

Fig. 43

On LL at a distance from the left index equal to $\log(\log A)$ is marked the number A, and on C at a distance from the left index equal to $\log B$ is marked the number B.

The sum of these two lengths is $\log(\log A) + \log B$. But according to (49) this is equal to $\log(\log x)$. If therefore the left index of C is set at A on LL as shown in Fig. 43, then x on LL will be at B on C. Thus to solve (49), which is (46), set the C index to A on LL and at B on C read x on LL. As an example consider the equation (46a) of Article 30, which has already been solved in three settings without the LL scale,

$$13.65^{2.37} = x. \qquad (46a)$$

It is solved on the log-log rule in one setting as follows:

(1) To 13.65 on LL set C index,
(2) To 2.37 on C set runner,
(3) At indicator on LL read 186.

This setting is diagrammed in Fig. 44.

C	Set 1	R to 2.37
LL3	To 13.65	Read 1.86

Fig. 44

Consider next equation (47),

$$x^A = B. \qquad (47)$$

Taking logarithms as in the last case

$$A(\log x) = (\log B),$$

and again taking logarithms,

$$\log A + \log(\log x) = \log(\log B),$$

or $\qquad \log(\log B) - \log A = \log(\log x), \qquad (50)$

USE OF THE LOG-LOG SCALES 125

which is logarithmically the same as (49) or (48). The solution is shown in Fig. 45 (a).

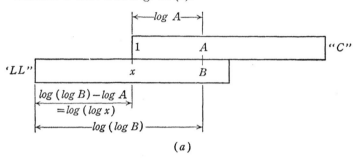

(a)

C	Set 3.62	R to 1
LL3	To 79.5	Read 3.35

(b)

Fig. 45

By tracing the steps through as was done above with Figs. 41 and 43 it is seen that A on C is set at B on LL and x is found on LL at the C index. As an example of this solution there is diagrammed in Fig. 45 (b) the setting for equation (47a) which is stated as follows:

(1) To 79.5 on LL set 3.62 on C,
(2) To C index set runner,
(3) At indicator on LL read 3.35.

For comparing and summarizing the above results and settings we write here the three equations (43), (46), (47):

$$b^x = a, \quad A^B = x, \quad x^A = B.$$

In this form it is seen that the three equations are all exponential in form with the unknown quantity x in a different position in each and with two known quantities in each, the first equation being the form ordinarily referred to as the "exponential equation." As we have seen above from the logarithmic forms these three equations all involve the same type of setting on the C and LL scales, one number, the exponent, being always read on C and the base and power on LL, the difference being that the steps are taken in different sequences in the three settings. If we take the Ath root of the last equation it is seen at once that it may be written as $x = \sqrt[A]{B}$. The three equations are therefore

(i) $x = A^B$,   (ii) $x = \sqrt[A]{B}$,   (iii) $b^x = a$.

In this form it is seen that (i) represents the Bth power of the number A; (ii) represents the Ath root of B; and (iii) is an exponential equation.

    In (i) A is a *number*, B is the *exponent* of the power, and x is the Bth power.
    In (ii) B is a *number*, A is the *index* of the root, and x is the Ath *root*.
    In (iii) b is the *base* number, x is the *exponential*, and a is the *power*.

These terms will be used below in stating rules for the slide rule calculation of x in (i), (ii), (iii).

Now a, b, A, B may be any numbers whatever; since all

USE OF THE LOG-LOG SCALES   127

three of the above equations have been solved in the settings explained and illustrated above, it is seen therefore that by means of the C and LL1, LL2, LL3 and the B and LL0 scales any root or power of any number (within the range of the scales) may be found, and exponential equations may be solved, at one setting of the slide and runner. Referring back to the settings corresponding to (i), (ii), (iii) in Figs. (43), (45a), (41) respectively, we have therefore the following log-log settings for powers, roots and exponentials:

(i) To find any POWER of any number:

(1) To the *number* on LL set C index,
(2) To the power *exponent* on C set runner,
(3) At indicator on LL read the *power*.

(ii) To find any ROOT of any number:

(1) To the *number* on LL set root *index* on C,
(2) To C index set runner,
(3) At indicator on LL read the *root*.

(iii) To solve an EXPONENTIAL equation:

(1) To the *base* on LL set C index,
(2) To the *power* on LL set runner,
(3) At indicator on C read the *exponential*.

The numerical examples so far solved have required only one part of the log-log scale, LL3, and the C scale as used with LL3 has been used to read from 1 to 10, so that the numbers printed on C were units. Let us now try to solve the exponential equation (42) in Article 29 by the setting (iii) above. Thus

(1) To 4.76 on LL set C index,
(2) To 2.19 on LL set runner,
(3) At indicator on C read .502.

Step (1) is made at once but when the attempt is made to set the runner on 2.19 on LL, 2.19 is not found on LL3. Since LL3 begins at $e = 2.7183$ the number 2.19 is found on the next lower range or section, LL2. With the runner at 2.19 on LL2 the reading for x on C is 502 but if the step-by-step logarithmic solution in Article 29 is inspected it will be seen that the solution is .502 and not 5.02 as in our previous readings of C in connection with LL3.

Similarly let us find the fifth root of the number 1.25, that is, find $x = \sqrt[5]{1.25}$. According to setting (ii) above the root index 5 on C is set to 1.25 on LL2, the runner is then set on the C index and x is to be read at the indicator on LL. At the indicator on LL3, however, is 87 which is obviously not $\sqrt[5]{1.25}$. At the indicator on LL2 is 1.562 which also is obviously not $\sqrt[5]{1.25}$. If, however, the LL1 reading, 1.0456, is raised to the 5th power by the ordinary logarithmic method the result is 1.25. Therefore $x = \sqrt[5]{1.25} = 1.0456$ is to be read on LL1.

As a third example let us make the setting for $x = 2^7 = 128$. According to setting (i) above the (right) C index

USE OF THE LOG-LOG SCALES 129

is set at 2 on LL2 and the runner at 7 on C, and x is read at the indicator on LL. Since 2^7 is greater than 2 the result is to be sought farther along to the right on LL and when the end of LL2 is reached it is to be taken as continuing from 2.7183 at the left index of LL3. Thus 128 is found at the indicator on LL3.

A single setting will bring out the different relations of C to the three parts of LL. Thus let us raise 4 to the powers 5, .5, and .05. In accordance with setting (i) the C index is set at 4 on LL3 and the runner at the figure 5 on C. Now $4^5 = 1024$, $4^{.5} = 4^{1/2} = \sqrt{4} = 2$, $4^{.05} = 4^{1/20} = \sqrt[20]{4} = 1.0718$. With the runner at the figure 5 on C the numbers read at the indicator on LL (according to (3) of setting (i) above) are 1024 on LL3, 2 on LL2, and 1.0718 on LL1.

The results of the four examples just discussed indicate that the numbers on C are read as units in connection with LL3, they are read as tenths in connection with LL2, and as hundredths in connection with LL1. This might have been stated directly without examination by recalling the limits and range of each of the sections of LL. Thus LL1 runs from the .01 to the .10 powers of e (the hundredths); LL2 from 0.1 to 1.0 (tenths) powers; and LL3 1 to 10 (units) powers. Also, as found in the preceding examples, when a reading cannot be found on that section of LL on which the setting was begun it will be found on

130 MODIFIED FORMS OF THE MANNHEIM RULE

a higher or lower section which is to be considered as a direct continuation of the complete LL scale. Since the decimal point is given on LL its position on C in any setting of C with LL will be obvious on inspection of the example.

Summarizing the preceding examples and the discussion of the last paragraph we can make the general statement (refer also to Article 31 and Fig. 40): LL1, LL2, LL3 read from $e^{.01}$ to $e^{.1}$ to e^1 to e^{10} and when slide and stock indexes are aligned and C is referred to LL3, LL2, LL1 the numbers on C are units, tenths, hundredths, respectively.

From this statement it is seen at once that there is another way of viewing the LL scale. Thus with slide and stock indexes aligned, 1 on C is at e^1 on LL3 and 10 on C is at e^{10} on LL3; similarly 4 on C is at e^4 on LL3, and so for any other number on C: that power of e is at that point on LL3. Similarly .1 on C is at $e^{.1}$ on LL2 and 1 on C is at e^1 on LL2, with corresponding relations for the intermediate numbers, and for the LL1 indexes and intermediate values. Thus if the runner is set at any number from .01 to .1 to 1 to 10 on C that power of e is read at the indicator on the corresponding section of LL with the decimal point placed. The LL scale, therefore, furnishes a table of exponentials such as is usually given in engineering and mathematical handbooks. Sections 1, 2, 3 of LL cover the range $e^{.01}$ to e^{10} of such a table and the negative exponentials $e^{-.1}$ to e^{-3} are given on LL0 which is used in connection with A or B in the same manner as the other three sections are used with C.

The further use of LL0 in connection with A will be understood from the following considerations. LL0 covers the range $e^{-.03} = .9703$ to $e^{-3} = .0498$, as stated in Article 31 (to three figures .0498 was there given as .050). $e^{-.1} = .9048$ is at both the left and right indexes and $e^{-1} = .3679$ at the middle index, while both $e^{-.03}$ and e^{-3} are at the point 3 on A2. The figure 3 on A2 is therefore the beginning and the end of LL0. When A is read in direct connection with LL0 therefore the figure 3 on A2 is to be read as .03 at the beginning of LL0 and the remaining figures on A2 as .04 to .1; A1 is to be read as .1 to 1, and the first part of A2 as 1 to 3. A minus sign $(-)$ is to be placed before each of these A numbers because the exponentials here are negative.

The complete table of exponentials furnished by LL in connection with A and C therefore covers the range $e^{-.03}$ to e^{-3} and $e^{.01}$ to e^{10}, which is ample for all practical purposes.

Now in accordance with the definition of logarithms to any base in Article 9 and of natural logarithms (base e) in Article 29 the exponents in the table of exponentials are nothing more than the natural logarithms of the numbers corresponding to them. Thus $e^{10} = 22026.3$

132 MODIFIED FORMS OF THE MANNHEIN RULE

and $\log_e 22026.3 = 10$ according to equations (14). Similarly $e^{-.03} = .9703$ and $\log_e .9703 = -.03$, which is in accordance with formula (41), decimal fractions having negative logarithms. Therefore if the runner is set at any number on LL its natural logarithm is found at the indicator on A or C (indexes being aligned). The characteristic of the natural logarithm is included in the A or C scale reading and the decimal point is placed as already explained in connection with the reading of those scales with LL. The LL scale (complete) when thus used in connection with A and C furnishes a direct reading table of natural logarithms of numbers from .0498 to .9705 and 1.0101 to 22026. The gap 0.9705 to 1.0101 exists because $\log_e 1 = 0$ and the A and C scales do not extend to zero.

By making use of the logarithm conversion formulas (35) and (36) the logarithm of numbers to any base can be found by setting the C index to the *base* on LL; if the runner is then set to any number on LL its logarithm to the chosen base is read at the indicator on C. The reading so found includes the characteristic and the decimal point is to be placed in accordance with the rules of reading given above. In this way the common logarithms can be found by setting the C index to 10 on LL. Since scale C can ordinarily be read to three figures, however, while the mantissa of the common logarithm can be read to three figures on scale L, it is better to read common logarithms on L and add the characteristic by the usual rule.

Numbers read on A, C and LL in any of the methods discussed in this article may of course then be used in connection with any of the other scales on the log-log slide rule as already explained in connection with the use of those scales on the Mannheim, polyphase and duplex rules. These various uses will not be described here but a large number of typical settings for the various scales of the log-log duplex rule are given in the next chapter.

D. THE SLIDE RULE IN TRIGONOMETRY

33. Introductory. — In Articles 22 and 23 the S and T scales of the slide rule are described and their use in finding the trigonometric functions of angles is explained, as is their use in multiplying other numbers by these functions. By applying these operations to the sides and angles of triangles all the triangle solutions of plane trigonometry may be carried out on the slide rule. With the exception of one or two cases no single computation in any of these solutions requires more than one setting of slide and runner.

In the following articles the operations involved in these solutions will be described and explained. In order to perform these operations on the Mannheim and polyphase slide rules it is necessary to reverse (but not invert) the slide so as to have the S scale contiguous to A and the T scale to D. When this is done these four scales may by means of the runner be used together in the same manner on all slide rules; the descriptions given below apply therefore to all the forms already described.

In Article 19 two special gauge points (π and $\frac{1}{4}\pi$) on the A and B scales were described and their uses explained there and elsewhere. A few operations involving these gauge points which are useful in trigonometry will be given in the next article.

34. Radian and Degree Angle Measure Conversion. — Since π radians equal 180 degrees, if any angle be expressed in both radians and degrees then π: 180 :: (no. rad.): (no. deg.), or

$$\frac{\pi}{180} = \frac{\text{radians}}{\text{degrees}} \quad \begin{matrix}(A)\\(B)\end{matrix} \qquad (51)$$

the capital letters at the right in parentheses indicating the scales on which the proportion is to be set up. Therefore to convert degrees to radians or radians to degrees we have the following operation:

$$\begin{cases}(1) \text{ To } \pi \text{ on A set 180 on B, and}\\(2) \text{ At radians on A read degrees on B, or}\\(3) \text{ At degrees on B read radians on A.}\end{cases}$$

If it is desired to convert half a given angle from radians to degrees the formula

$$\frac{\pi}{90} = \frac{\text{radians in angle}}{\text{degrees in half angle}} \quad \begin{matrix}(A)\\(B)\end{matrix} \qquad (52)$$

SOLUTION OF RIGHT TRIANGLES 135

is used. It is set up on A, B as indicated. In order to convert an angle in degrees to half the angle in radians the formula is:

$$\frac{\pi}{360} = \frac{\text{radians in half angle} \quad (A)}{\text{degrees in angle} \quad (B)} \tag{53}$$

The above conversions can be more accurately but less conveniently carried out on the C, D scales by making use of the following three corresponding proportions and setting them up on C, D in the usual manner of proportions:

$$\frac{134}{2.34} = \frac{\text{degrees}}{\text{radians}} \tag{51a}$$

$$\frac{134}{4.68} = \frac{\text{degrees in half angle}}{\text{radians in angle}} \tag{52a}$$

$$\frac{134}{1.17} = \frac{\text{degrees in angle}}{\text{radians in half angle}} \tag{53a}$$

35. Solution of Right Triangles. — If the standard notation for right triangles given in Fig. 46 is used the fundamental formulas are:

$$\frac{a}{c} = \sin A = \cos B$$

$$\frac{b}{c} = \cos A = \sin B$$

$$\frac{a}{b} = \tan A = \cot B$$

also,
$$\frac{a}{\sin A} = \frac{b}{\sin B}.$$

Since the angle C is in every case equal to 90° it is not necessary to state it in every solution.

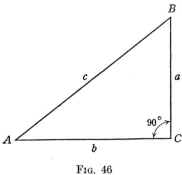

Fig. 46

For purposes of slide rule settings it is convenient to write the fundamental formulas in the form of proportions. Thus the first one above may be written $\frac{a}{\sin A} = \frac{c}{1}$ but $1 = \sin 90°$. Therefore $\frac{a}{\sin A} = \frac{c}{\sin 90}$, which may be set up on the A, S scales exactly as written. Similarly the third of the above formulas may be written $\frac{\tan A}{a} = \frac{\tan 45}{b}$ since $\tan 45° = 1$, and in this form the proportion may be set up directly on the T, D scales. To find either angle A or B when the other is known the relation $A + B = 90°$ is used and the known angle is subtracted from 90° to give the other.

In each case of triangle solution the two given parts,

SOLUTION OF RIGHT TRIANGLES

the three required parts, the appropriate solution formulas and the slide rule setting will be stated and a numerical example will be diagrammed. The decimal point will be placed as explained in Articles 22 and 23, or to correspond with that in the corresponding member of the solution proportion.

If in any case it is required to use the tangent of an angle less than $5° 43'$ the S scale may be used instead of T as explained in Article 23.

CASE I. — Given two legs a, b; to find A, B, c.

$$\frac{1}{b} = \frac{\tan A}{a}, \quad B = 90 - A, \quad \frac{a}{\sin A} = \frac{c}{1}.$$

(i) $\begin{cases} (1) \text{ To } b \text{ on D set } 45° \text{ on T,} \\ (2) \text{ To } a \text{ on D set runner,} \\ (3) \text{ At indicator on T read } A. \end{cases}$

(ii) $\begin{cases} (1) \text{ To } a \text{ on A set } B \text{ on S,} \\ (2) \text{ To 90 on S set runner,} \\ (3) \text{ At indicator on A read } c. \end{cases}$

Example: $a = .0227$, $b = .084$

T	Set 45°	$A = 15° 7'$
D	To 84	R to .0227
	(i)	

A	To .0227	$c = .087$
S	Set 15° 7'	R to 90°
	(ii)	

$$B = 90° - 15° 7' = 74° 53'$$

As an example of the use of S instead of T let $a = 64.4$, $b = 5.99$, find B by setting (i)

A	To 64.4	R to 5.99
S	Set 90°	$B = 5° 19'$
	(i)	

138 MODIFIED FORMS OF THE MANNHEIM RULE

Case II. — Given leg and hypotenuse b, c; to find a, A, B.
$$\frac{c}{1} = \frac{b}{\sin B}, \quad A = 90 - B, \quad \frac{b}{\sin B} = \frac{a}{\sin A}$$

(1) To c on A set 90 on S
(2) To b on A set runner
(3) At indicator on S read B; $A = 90 - B$, then
(4) To A on S set runner
(5) At indicator on A read a

Example: $b = .647$, $c = 7.46$.

A	To 7.46	R to .647	$a = 7.43$
S	Set 90°	$B = 4° 59'$	R to $(90 - 4° 59') = 85° 1'$

Case III. — Given leg and adjacent angle b, A; to find B, c, a.
$$B = 90° - A, \quad \frac{b}{\sin B} = \frac{c}{1} = \frac{a}{\sin A}$$

After subtracting A from 90° to find B then

(1) To b on A set B on S
(2) To 90 on S set runner
(3) At indicator on A read c; also
(4) To A on S set runner
(5) At indicator on A read a

Example: $b = .084$, $A = 15° 9'$. $B = 90° - 15° 9' = 74° 51'$

A	To .084	$c = .087$	$a = .0227$
S	Set 74° 51'	R to 90°	R to 15° 9'

SOLUTION OF RIGHT TRIANGLES   139

Case IV. — Given a leg and opposite angle: a, A; to find B, b, c.

$$B = 90° - A, \quad \frac{a}{\sin A} = \frac{b}{\sin B} = \frac{c}{1}$$

After subtracting A from 90° to find B, then

(1) To a on A set A on S
(2) To B on S set runner
(3) At indicator on A read b; also
(4) To 90° on S set runner
(5) At indicator on A read c.

Example: $a = 3.73, A = 24° 20'$. $B = 90° - 24° 20' = 65° 40'$

A	To 3.73	$b = 8.25$	$c = 9.05$
S	Set 24° 20'	R to 65° 40'	R to 90°

Case V. — Given hypotenuse and an acute angle: c, A; to find B, a, b

$$B = 90° - A, \quad \frac{c}{1} = \frac{a}{\sin A} = \frac{b}{\sin B}$$

After subtracting A from 90° to find B, then

(1) To c on A set 90° on S
(2) To A on B set runner
(3) At indicator on A read a; also
(4) To B on S set runner
(5) At indicator on A read b.

Example: $c = 426, A = 34° 15'$. $B = 90° - 34° 15' = 55° 45'$

A	To 426	$a = 240$	$b = 352$
S	Set 90°	R to 34° 15'	R to 55° 45'

If it is desired to read b in Case V, c in Case IV, a in Case III and a in Case II more accurately but with an extra setting the tangent formula may be used and the setting made on the T and D scales. Thus in the last example, Case V, $\dfrac{\tan A}{a} = \dfrac{1}{b}$. The setting is below.

T	Set 34° 15′	R to 45°
D	To 240	$b = 352$

36. Solution of Oblique Triangles. — Using the notation of Fig. 47 the direct solution formulas for the oblique triangle are

$$\frac{a}{\sin A} = \frac{b}{\sin B} = \frac{c}{\sin C};$$

$$\frac{\tan[\tfrac{1}{2}(A - B)]}{a - b} = \frac{\tan[\tfrac{1}{2}(A + B)]}{a + b} = \frac{\cot \tfrac{1}{2} C}{a + b},$$

with corresponding formulas for angles A, C and B, C.

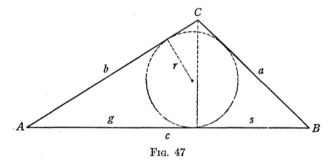

Fig. 47

When the sides are given two indirect methods may be used. One requires finding the radius r of the inscribed

SOLUTION OF OBLIQUE TRIANGLES

circle and the other requires finding the two segments g, s into which the longest side is divided by the perpendicular from the opposite vertex. In the first method when

$$S = \tfrac{1}{2}(a + b + c), \quad r = \sqrt{\frac{(S-a)(S-b)(S-c)}{S}}$$

$$\frac{\tan \tfrac{1}{2} A}{r} = \frac{1}{S-a}, \quad \frac{\tan \tfrac{1}{2} B}{r} = \frac{1}{S-b}, \quad C = 180° - (A+B).$$

In the second method (when the lengths are in the order of Fig. 47)

$$d = \frac{(b+a)(b-a)}{c}, \quad g = \tfrac{1}{2}(c+d), \quad s = \tfrac{1}{2}(c-d),$$

$$\frac{b}{1} = \frac{g}{\sin(90° - A)}, \quad \frac{a}{1} = \frac{s}{\sin(90° - B)}, \quad C = 180° - (A+B).$$

As in the solution of right triangles, there are five cases of oblique triangles to consider. The difference lies in the fact that in right triangles one angle is always known, which is not the case in oblique triangles.

CASE I. — Given a side and two adjacent angles: a, B, C; to find A, b, c.

$$A = 180° - (B + C), \quad \frac{a}{\sin A} = \frac{b}{\sin B} = \frac{c}{\sin C}.$$

After finding A by subtracting $(B + C)$ from 180°, then

(1) To a on A set A on S,
(2) At B on S read b on A,
(3) At C on S read c on A.

Example: $a = 25$, $B = 50°$, $C = 60°$;
$A = 180° - (50° + 60°) = 70°$.

A	To 25	$b = 20.6$	$c = 23.2$
S	Set 70°	R to 50°	R to 60°

CASE II. — Given two angles and a side opposite one of them: A, B, a; to find C, b, c.

$$C = 180° - (A + B), \quad \frac{a}{\sin A} = \frac{b}{\sin B} = \frac{c}{\sin C}.$$

After subtracting $(A + B)$ from 180° to find C the setting is the same as in Case I.

CASE III. — Given two sides and an angle opposite one of them: a, b, A; to find C, B, c.

$$\frac{a}{\sin A} = \frac{b}{\sin B}, \quad C = 180° - (A + B), \quad \frac{a}{\sin A} = \frac{c}{\sin C}.$$

(1) To a on A set A on S,
(2) At b on A read B on S; then $C = 180° - (A + B)$, and
(3) At C on S read c on A.

If A is acute and $a > b$ there is one solution; otherwise there are two solutions.

Example 1: One solution: $A = 32° 43'$, $a = .988$, $b = .672$.

A	To .988	R to .672	$c = 1.480$
S	Set 32° 43'	$B = 21° 34'$	R to 54° 17'

SOLUTION OF OBLIQUE TRIANGLES 143

Example 2: Two solutions: $A = 48° 34'$, $a = 46.24$, $b = 60$.

A	To 46.24	R to 60	$c_1 = 50.4$	$c_2 = 29.1$
S	Set 48° 34'	$B_1 = 76° 42'$ $B_2 = 180° - B$ $= 103° 18'$	R to $C_1 = 54° 44'$ $= 180° - (A + B_1)$	R to $C_2 = 28°8'$ $= 180° - (A + B_2)$

CASE IV. — Given two sides and the included angle: A, b, c; to find, B, C, a.

$$\tfrac{1}{2}(B + C) = 90° - \tfrac{1}{2} A, \quad \frac{\tan[\tfrac{1}{2}(B + C)]}{b + c} = \frac{\tan[\tfrac{1}{2}(B - C)]}{b - c},$$

$$B = [\tfrac{1}{2}(B + C)] + [\tfrac{1}{2}(B - C)], \quad C = [\tfrac{1}{2}(B + C)] - [\tfrac{1}{2}(B - C)],$$

$$\frac{b}{\sin B} = \frac{a}{\sin A}$$

(i) $\begin{cases} (1) \text{ To } (b + c) \text{ on D set } \tfrac{1}{2}(B + C) \text{ on T,} \\ (2) \text{ To } (b - c) \text{ on D set runner,} \\ (3) \text{ At indicator on T read } \tfrac{1}{2}(B - C). \end{cases}$

Then after finding B on C from the formulas,

(ii) $\begin{cases} (1) \text{ To } b \text{ on A set } B \text{ on S,} \\ (2) \text{ At } A \text{ on S read } a \text{ on A.} \end{cases}$

If A is obtuse operation (i) requires a single setting; if A is acute a double setting is required

Example 1: Single setting: $A = 110° 32'$, $b = 6.24$, $c = 2.35$.

$$[\tfrac{1}{2}(B + C)] = 90° - 55° 16' = 34° 44', \quad b + c = 8.59.$$
$$b - c = 3.89$$

(i)

T	Set 34° 44'	$[\tfrac{1}{2}(B - C)] = 17° 26'$
D	To 8.59	R to 3.89

$B = 34° 44' + 17° 26'$ $\quad C = 34° 44' - 17° 26'$
$ = 52° 10'$ $\quad = 17° 18'$

144 MODIFIED FORMS OF THE MANNHEIM RULE

(ii)
	A	To 6.24	$a = 7.39$
	S	Set 52° 10'	R to 180° − 110° 32' = 69° 28'

Example 2: Double setting: $A = 59° 13'$, $b = 226.2$, $c = 138.7$

$$[\tfrac{1}{2}(B + C)] = 90° - 29° 37' = 60° 23', \quad b + c = 364.9,$$
$$b - c = 87.5$$

(i)
	T	Set 45°	R to 29° 37'	Index to R	$[\tfrac{1}{2}(B - C)] = 22° 54'$
	D	To 364.9			R to 87.5

$$B = 60° 23' + 22° 54' \qquad C = 60° 23' - 22° 54'$$
$$= 83° 17' \qquad\qquad = 37° 29'$$

(ii)
	A	To 226	$a = 196$
	S	Set 83° 17'	R to 59° 13'

CASE V. — Given the three sides: a, b, c; to find A, B, C:

Method 1. — $S = \tfrac{1}{2}(a+b+c)$, $r = \sqrt{\dfrac{(S-a)(S-b)(S-c)}{S}}$

$$\frac{1}{S-a} = \frac{\tan \tfrac{1}{2} A}{r}, \quad \frac{1}{S-b} = \frac{\tan \tfrac{1}{2} B}{r}, \quad C = 180° - (A + B)$$

(i)
- (1) To $(S - a)$ on A set S on B,
- (2) To $(S - b)$ on B set runner,
- (3) To runner set B index,
- (4) To $(S - c)$ on B set runner,
- (5) At indicator on D read r.

(ii)
- (1) To $(S - a)$ on D set 45° on T,
- (2) At r on D read $\tfrac{1}{2} A$ on T.

(iii)
- (1) To $(S - b)$ on D set 45° on T,
- (2) At r on D read $\tfrac{1}{2} B$ on T.

From 180° subtract $(A + B)$ to find C.

SOLUTION OF OBLIQUE TRIANGLES 145

Example: $a = 260, b = 280, c = 300; S = 420, S - a = 160, S - b = 140, S - c = 120.$

(i)
A	To 160			
B	Set 420	R to 140	Index to R	R to 120
D				$r = 8$

(ii) $\dfrac{T}{D} \Big| \dfrac{\text{Set left I.}}{\text{To 160}} \Big| \dfrac{\tfrac{1}{2}A = 26° 35'}{\text{R to 8}}$ (iii) $\dfrac{T}{D} \Big| \dfrac{\text{Set L.I.}}{\text{To 140}} \Big| \dfrac{\tfrac{1}{2}B = 29° 45'}{\text{R to 8}}$

$A = 53° 10',\quad B = 59° 30',\quad C = 180° - 112° 40'$
$$= 67° 20'$$

Method 2. — $d = \dfrac{(b+a)(b-a)}{c},\ g = \tfrac{1}{2}(c+d),\ s = \tfrac{1}{2}(c-d)$

$$\dfrac{b}{1} = \dfrac{g}{\sin(90° - A)},\quad \dfrac{a}{1} = \dfrac{s}{\sin(90° - B)},$$

$C = 180° - (A + B).$

(i) $\begin{cases} (1)\ \text{To } (b+a) \text{ on D set } c \text{ on C,} \\ (2)\ \text{At } (b-a) \text{ on C read } d \text{ on D.} \end{cases}$

(ii) $\begin{cases} (1)\ \text{To } b \text{ on A set } 90° \text{ on S,} \\ (2)\ \text{At } g \text{ on A read } (90° - A) \text{ on S.} \end{cases}$

(iii) $\begin{cases} (1)\ \text{To } a \text{ on A set } 90° \text{ on S,} \\ (2)\ \text{At } s \text{ on A read } (90° - B) \text{ on S.} \end{cases}$

Example: $a = 260,\ b = 280,\ c = 300,\ (b+a) = 540,\ (b-a) = 20.$

(i) $\dfrac{C}{D} \Big| \dfrac{\text{Set 300}}{\text{To 540}} \Big| \dfrac{\text{R to 20}}{d = 36}$ $\quad g = \tfrac{1}{2}(300 + 36) = 168$
$\quad s = \tfrac{1}{2}(300 - 36) = 132$

$\dfrac{A}{S} \Big| \dfrac{\text{To 280}}{\text{Set 90°}} \Big| \dfrac{\text{R to 168}}{(90° - A) = 36° 50'}$ (ii) $\quad \dfrac{A}{S} \Big| \dfrac{\text{To 260}}{\text{Set 90°}} \Big| \dfrac{\text{R to 132}}{(90° - B) = 30° 30'}$ (iii) $\quad q\tfrac{1}{2}$

$A = 53° 10',\quad B = 59° 30',\quad C = 180° - 112° 40'$
$$= 67° 20'$$

146   MODIFIED FORMS OF THE MANNHEIM RULE

This completes the solution of all cases of plane triangles. Problems in spherical trigonometry will not be treated here.

37. Slide Rule Check of Logarithmic Solutions. — So far we have only considered the slide rule as performing original calculations. It is, of course, also very useful in checking calculations made more accurately by other methods, such as "long hand" arithmetical calculations or logarithmic computations, and as an aid in carrying out logarithmic computations themselves. One of the most useful and convenient of these checks is that of triangle solutions carried out with logarithm tables.

In any plane triangle, right or oblique, the so-called sine law must hold:

$$\frac{a}{\sin A} = \frac{b}{\sin B} = \frac{c}{\sin C},$$

where A, B, C are the angles and a, b, c the corresponding opposite sides. In the usual notation (as used in Article 35) for right triangles C is $90°$ but the equations still apply. If therefore values of all the six parts of any triangle are known and any angle is set on the S scale at its opposite side on A, then each of the other two angles on S will be at its opposite side on A. Therefore to check any logarithmic solution or graphic solution of a triangle set any one of the angles obtained on S to its side on A.

If the solution is correct then each of the other two angles on S will be at its opposite side on A, thus:

$$\frac{A}{S} \parallel \frac{a}{A} \mid \frac{b}{B} \mid \frac{c}{C}$$

If they are not so found, the solution is not correct; if they are so found, the solution is correct at least to the first few figures of each side and to within a very few minutes of each angle.

38. Interpolation of Logarithms. — If a table of logarithms is not provided with auxiliary tables of proportional parts interpolation between tabulated values is tedious and liable to error. If such auxiliary tables are provided they frequently require interpolation in order to obtain the last figure given by the table. The slide rule is simply and easily used to replace the table of proportional parts in the first case and to extend or supplement it in the second case.

When the differences between successive tabulated values are small the number differences and the corresponding logarithm differences are proportional. Therefore when the differences between any two consecutive numbers and between their logarithms are known a difference in one corresponding to an intermediate difference in the other is found by solving a proportion with three

148 MODIFIED FORMS OF THE MANNHEIM RULE

known members. The slide rule is therefore especially adapted to finding the required difference.

Let n and N be two consecutive numbers in a table of logarithms, and l and L their respective logarithms. Let D be the logarithm difference $(L - l)$; the corresponding number difference is $(N - n) = 1$. If then d is a fractional difference intermediate between n and N, and X is the corresponding logarithm difference, we have

$$\frac{1}{D} = \frac{d}{X}, \quad \text{or} \quad X = Dd. \tag{51}$$

As in every case, therefore, the solution may be performed as a multiplication or as a proportion. In either case it is stated as follows:

(1) To *total log diff.* on D set C index,
(2) To *partial number diff.* on C set runner,
(3) At indicator on D read *partial log. diff.*

As an example, find in a six-place table the common logarithm of 7935.73. From the table

$$\log 7936 = 3.899602$$
$$\log 7935 = 3.899547$$

$D = 55$ (in last two places)
For 7935.73, $d = .73$
$X = 40.1$ (in last two places)
∴ $\log 7935.73 = 3.899587$

The slide rule setting for the interpolation is:

C	Set 1	R to .73
D	To 55	$X = 40.1$

INTERPOLATION OF LOGARITHMS 149

The same form of setting holds for second differences when a table of proportional parts is supplied. Thus to find from such a table log 232.5648, from the table

$$\log 232.6 = 2.366610$$
$$\log 232.5 = 2.366423$$

1st diff. = 187 (in last three places)
For next fig. 7 p.p. = 130.9
"   "   " 6 p.p. = 112.2

2nd diff. = 18.7
For next figs. 48, p.p. = 8.99 (by slide rule)
For figs. 648, p.p. = 112.2 + 8.99 = 121.19, or
 = 121 (in last three places)
Log 232.5648 = 2.366544

The slide rule setting for the second interpolation is:

C	Set 1	R to 48
D	To 18.7	Read 8.99

Of course, as in the first example given, the logarithm difference 121 could be found just as well by reading directly on the slide rule without reference to the table of proportional parts. Thus by formula (51)

$$\frac{1}{187} = \frac{648}{X} \quad \text{and} \quad X = 121.$$

For a six-place table, where the logarithm differences do not go beyond three figures, therefore, the tables of proportional parts are unnecessary and the slide rule may be used directly for the full difference. With tables of seven and more figures the slide rule may be used as in

150   MODIFIED FORMS OF THE MANNHEIM RULE

the second example above to extend the tables of proportional parts.

For interpolating antilogarithms the procedure given above is simply reversed by taking X as known in proportion (51) and d as unknown. Thus to find antilog 1.452893:

$$\text{antilog } 1.453012 = 28.38$$
$$\text{antilog } 1.452859 = 28.37$$

$$D = 153$$
$$\text{For antilog } 1.452893,\ X = 34,$$
$$d = .222 \text{ (next figs. after 37)}$$
$$\text{antilog } 1.452893 = 28.37222$$

The slide rule setting for this form of interpolation is:

C	Set 153	R to 1
D	To 34	$d = .222$

Natural logarithms and antilogarithms may be interpolated in the same way as for common logarithms. The natural and logarithmic angular functions may also be interpolated in the same way. Since the procedure for all these is precisely the same as that given above they will not be described here.

CHAPTER IV

TYPICAL PROBLEMS AND SLIDE RULE SETTINGS

A. INTRODUCTION.

The various forms of slide rule using the standard Mannheim scales being adapted to any form of computation involving multiplication, division, proportion, powers, roots, common and natural logarithms, exponentials and trigonometric functions are capable of performing an indefinite number and variety of operations. With any given type of slide rule and problems certain appropriate settings may be developed to suit individual needs. These may be modified, varied or adapted to suit other needs or scale arrangements. As an illustration of the development of such settings there is given below in Section B a list of diagrammed settings for a number of commercial and engineering problems which involve only the A, B, C and D scales. These are of course adapted to any of the standard rules. Lists of settings involving the ordinary and standard operations with these scales, such as those described in Chapter II, and their simple combinations will not be given.

In Section C are given the settings and gauge points

152 TYPICAL PROBLEMS AND SLIDE RULE SETTINGS

for a number of the more commonly useful geometrical mensuration calculations and for a large number of standard conversions from one to another of the various systems of units of measurement. The gauge points apply to the C and D scales; these settings may therefore be used with any of the standard forms of rules. Many of these conversion gauge points and settings are printed on the back of the stock of the Mannheim rule by most manufacturers.

The settings given in the remaining Sections of this chapter in the main apply particularly to other forms of rule than the simple Mannheim but where they involve only the A, B, C and D scales they may of course also be used on the Mannheim slide rule.

B. TYPICAL PROBLEMS AND SETTINGS FOR THE A, B, C AND D SCALES.

In the thirty-seven typical settings which follow the computing formula is given for many of the problems; many others involve the simple standard operations and the formula is not given with the setting. The method which has been used throughout this book and explained in Chapter II is used. Horizontal lines represent the edges of contiguous scales and the scales are indicated by capital letters above and below these lines in the positions in which they occur on the slide rule. Vertical lines

TYPICAL PROBLEMS AND SETTINGS   153

separate the successive steps in an operation; double vertical lines indicating that the slide is to be reset or removed and reversed or inverted, or serving to separate the letters used to indicate the scale from other letters involved in the setting of the problem. While these settings apply only to the A, B, C and D scales they indicate the method of working out many different types of problem and may serve as models of the methods of combining other scales than these four. Some of these operations may be more conveniently performed by means of other scales, as will appear in later sections devoted to the several types of rules.

1. — Diameters and Areas of Circles. $A = .7854 D^2$

A ‖	To 205			A	To 11	
B	Set 161	Find Areas		B	Set 6	Find Areas in square feet.
C			*or*	C		
D		Above Diameters		D		Above Diameters in inches

2. — To Calculate Selling Prices of Goods, with Percentage of Profit on Cost Price

C ‖	Set 100	Below cost price
D ‖	To 100 plus percentage of profit	Find selling price

3. — To Calculate Selling Prices of Goods, with Percentage of Profit on Selling Price

C ‖	Set 100 less percentage of profit	Below cost price
D ‖	To 100	Find selling price

154 TYPICAL PROBLEMS AND SLIDE RULE SETTINGS

Example: If goods cost 45 cents a yard, at what price must they be sold to realize 15 per cent profit on the selling price?

C	Set 85 (= 100 − 15)	Below 45
D	To 100	Find 53 — Answer

4. — To find the Area of a Ring. $A = \dfrac{(D + d) \times (D - d)}{1.2732}$

C	Set sum of the two diameters	Find area.
D	To 1.273	Above difference of the two diameters

5. — Compound Interest

Set the left index of C, to 100 plus the rate of interest, on D, then take the corresponding number on the scale of Equal Parts, and multiply it by the number of years. Set this product on the scale of E.P. to the index on the under side of the Rule, then on D will be found the amount of any coinciding sum on C for the given years at the given rate.

Example: Find the amount of $150 at 5% at the end of 10 years.

C	Set 100	E.P. = 21.2 × 10 = 212	212 to 1	C	Below $150
D	To 105	Under side of Rule	and Slide	D	Find $244.35 — Answer

We thus obtain on D, below 1 on C, a gauge-point for 10 years at 5%, and can obtain in like manner similar ones for any other number of years and rate of interest.

6. — Levers

C	Set distance from fulcrum to power or weight applied	Find power or weight transmitted
D	To distance from fulcrum to power or weight transmitted	Above power or weight applied

TYPICAL PROBLEMS AND SETTINGS 155

7. — Diameter of Pulleys or Teeth of Wheels

$$\frac{\text{in'd C}}{D} \left\| \begin{array}{l} \text{Set diameter or teeth of Driving} \\ \hline \text{To revolutions of Driving} \end{array} \right. \begin{array}{l} \text{Diameter or teeth of Driven} \\ \hline \text{Revolutions of Driven} \end{array}$$

or, $\dfrac{C}{D}$ ‖ $\begin{array}{l}\text{Set diameter or teeth of Driving} \\ \hline \text{To diameter or teeth of Driven}\end{array}$ $\begin{array}{l}\text{Revolutions of Driven} \\ \hline \text{Revolutions of Driving}\end{array}$

8. — Diameter of Two Wheels to Work at Given Velocities

C	Set distance between their centers	Find diameter
D	To half Sum of their revolutions	Above revolutions of each.

Example: A shaft makes 21 revolutions, and is to drive another shaft which should make 35 revolutions. The distance between their centres is 48 inches. What should be the diameters of the gears?

C	Set 48	Find 36	Find 60
D	To 28(= 21 + 35 ÷ 2)	Above 21	And above 35

The two wheels must therefore be 36 and 60 inches diameter.

9. — Strength of Teeth of Wheels $\quad P = \dfrac{\sqrt{H}}{0.6\ V}$

A	To H.P. to be transmitted		
B			
C	Set gauge-point 0.6	R to 1 Velocity in ft. per second to R	Under 1
D			Pitch in inches.

10. — Diameter and Pitch of Wheels $\quad N = \dfrac{D \times \pi}{P}$

C	Set 226	R to 1	Pitch to R	Under diameter of pitch circle
D	To 710			Find number of teeth

11. — Strength of Wrought Iron Shafting

$$D = \sqrt[3]{\frac{83\,H}{N}} \text{ for crank shafts and prime movers.}$$

$$D = \sqrt[3]{\frac{65\,H}{N}} \text{ for ordinary shafting.}$$

A	To 83 or 65		
B	Set revolutions per min.	R to Ind. H.P.	Number coinciding with R = diameter
C			Under 1
D			Same coinciding number = diameter

NOTE. — In this, as in other cases, the coefficients (83 and 65) may be altered to suit individual opinions, without in any way altering the methods of solution.

12. — To find the Change Wheel in a Screw-Cutting Lathe

$$N = T\frac{S \times W}{M \times P}$$

$$W = N\frac{M \times P}{T \times S}$$

Where
- N = Number of threads per inch to be cut
- T = Number of threads per inch on traverse screw.
- M = Number of teeth in wheel on mandril.
- W = Number of teeth in stud wheel (gearing in M).
- P = Number of teeth in stud pinion (gearing in S).
- S = Number of teeth in wheel on traverse screw.

C	Set T	R to P	S to R	Under M
D	To N			Find number of teeth in W or stud wheel.

13. — Rules for Good Leather Belting

$$W = \frac{600 \text{ or } 375 \text{ H.P.}}{V \text{ ft. per min.}}$$

C	Set 600	Find width in inches
D	To velocity in feet per min.	Above actual horse power

for SINGLE BELTS.

TYPICAL PROBLEMS AND SETTINGS 157

C	Set 375	Find width in inches	for DOUBLE BELTS.
D	To velocity in feet per min.	Above actual horse power	

14. — Best Manilla Rope Driving

A	To velocity in feet per min.	Find ACTUAL HORSE POWER
B	Set 307	
C		Above diameter in inches
D		

A	To 4	Find STRENGTH IN TONS
B		
C	Set 1	Above diameter in inches
D		

A	To 107	Find WORKING TENSION IN POUNDS
B		
C	Set 1	Above diameter in inches
D		

A	To 0.28	Find WEIGHT PER FOOT IN POUNDS
B		
C	Set 1	Above diameter in inches
D		

15. — Weight of Iron Bars in Pounds per Foot Length

A	To 1	Weight of SQUARE BARS
B	Set 3	
C		Above width of side in inches
D		

158 TYPICAL PROBLEMS AND SLIDE RULE SETTINGS

A	To 55	Weight of ROUND BARS
B	Set 21	
C		Above diameter in inches
D		

C	Set 0.3	Below thickness in inches
D	Breadth in inches	Weight of FLAT BARS

16. — Weight of Iron Plates in Pounds per Square Foot

C	Set 32	Below thickness in thirty-seconds of an inch
D	To 40	Find weight in pounds per square foot

17. — Weights of Other Metals

C	Set 1	Below G.P. for other metals
D	To weight in iron	Find weight in other metals

Gauge-points of other metals, and weight per cubic foot

	W. I.	C. I.	Cast Steel.	Steel Plates.	Copper.	Brass.	Lead.	Cast Zinc.
G. P.	1	.93	1.02	1.04	1.15	1.09	1.47	.92
Weight	480	450	490	500	550	525	710	440 pounds

Example: What is the weight of a bar of copper, 1 foot long, 3 inches broad and 2 inches thick?

C	Set 0.3	R to 2 inches thick	1 to R	Below G.P. 1.15
D	To 3 inches broad			Find 23 pounds — Answer

18. — Weight of Cast Iron Pipes

C	Set .4075	Below DIFFERENCE of inside and outside diameters in inches
D	To SUM of inside and outside diameters in inches	Find weight in pounds per lineal foot

G.P. for other metals	Brass.	Copper.	Lead.	W. Iron
	.355	.333	.259	.38

TYPICAL PROBLEMS AND SETTINGS

19. — Safe Load on Chains

A		Safe load in tons
B	Set 36	Above 1
C		
D	To diameter in sixteenths of an inch	

20. — Gravity

C	Set 1	Below 32.2
D	To seconds	Velocity in feet per second

A	Space fallen through in feet	
B		
C	Set 1	Under 8
D		Velocity in feet per second

A		Space fallen through in feet
B		Above 16.1
C	Set 1	
D	To seconds	

21. — Oscillations of Pendulums

A		
B	Set length pendulum in inches	
C		Below 1
D	To 375	Number oscillations per minute

22. — Comparison of Thermometers

C	Set 5	Degrees CENTIGRADE
D	To 9	Degrees + 32 = FAHRENHEIT

160 TYPICAL PROBLEMS AND SLIDE RULE SETTINGS

C	Set 4	Degrees REAUMUR
D	To 9	Degrees + 32 = FAHRENHEIT
C	Set 4	Degrees REAUMUR
D	To 5	Degrees CENTIGRADE

23. — Force of Wind

A		
B	Set 10	Find pressure in pounds per square foot
C		
D	To 66	Velocity in FEET *per second*
A		
B	Set 10	Find pressure in pounds per square foot
C		
D	To 45	Velocity in MILES *per hour*

24. — Discharge from Pumps

A		Gallons delivered per stroke
B	Set 29.4	Stroke in feet
C		
D	To diameter in inches	

25. — Diameter of Single-acting Pumps

A	To 29.4			
B	Set length stroke in ft.	R to gallons to be delivered per min.	No. strokes per min. to R	
C				Below 1
D				Diam. pump in inches

TYPICAL PROBLEMS AND SETTINGS 161

26. — Horse Power Required for Pumps

C	Set G.P.	Height in feet to which the water is to be raised
D	To cubic feet or gallons to be raised per minute	Horse power required

Gauge Points with different Percentages of Allowance

Per cent	None	10	20	30	40	50	60	70	80
For Gallons, Imp.	3300	3000	2750	2540	2360	2200	2060	1940	1835
For C. Feet	528	480	440	406	377	352	330	311	294
For U. S. Gallons	3960	3600	3300	3050	2830	2640	2470	2330	2200

27. — Theoretic Velocity of Water for any Head

A	Head in feet	
B		
C	Set 1	Under 8
D		Velocity in feet per second

28. — Theoretical Discharge from an Orifice 1 inch Square

A			
B	Set 1	Under head in feet	If the hole is round and 1 inch diameter, the G.P. is 2.62.
C			
D	To G.P. 3.34	Discharge in cubic feet	

29. — Real Discharge from Orifice in a Tank, 1 inch Square

A			
B	Set 1	Under head in feet	If the hole is round and 1 inch diameter, the G.P. is 1.65.
C			
D	To 2.1 G.P.	Discharge in cubic feet per minute with coefficient .63.	

162   TYPICAL PROBLEMS AND SLIDE RULE SETTINGS

Gauge Points for Other Coefficients

Coefficient	.60	.66	.69	.72	.75	.78	.81	.84	.87	.90	.93	.96
G. P. SQUARE,	2.	2.2	2.3	2.4	2.5	2.6	2.7	2.8	2.9	3.	3.1	3.2
G. P. ROUND	1.57	1.73	1.80	1.88	1.96	2.04	2.12	2.20	2.28	2.36	2.44	2.52

30. — Discharge from Pipes When Real Velocity Is Known

Inverted
A		Discharge in cubic feet per minute.
C		Above 1.75
B	Velocity in feet per second	
D	Diameter in inches	

31. — Delivery of Water from Pipes $W = 4.71 \sqrt{\dfrac{D^5 H}{L}}$

C	Set 1	E P × 5 = x	x to I	Eytelwein's Rule.
D	To diameter in inches	Under side of Rule		(Con't. to next line.)

A				To Z	
B				Set 1	
C	R to 1	Length in feet to R	Under head in feet		Under 4.71
D			Find Z		Cubic ft. per min.

This is worked out similarly to Formula 5, which is explained in full.

32. — Gauging Water with a Weir

Inverted
A		
C		
B	Depth in inches	Under 4.3
D	Depth in inches	Discharge in cubic feet per minute from each foot width of sill.

TYPICAL PROBLEMS AND SETTINGS   163

33. — Discharge of a Turbine $\dfrac{\sqrt{H \times V}}{0.3} = D$

A			
Inverted { C			Under 0.3
B		Head in feet	
D		Square inches water vented	Cubic feet discharged per minute.

34. — Revolutions of a Turbine

A	To head in feet			
B C	Set diameter in inches	R to 1840	1 to R	Under rate of peripheral velocity
D				Find revolutions per minute.

35. — Horse Power of a Turbine

C	Set 530	R to discharge per c. ft. per min.		1 to R	Percentage useful effect
D	Head in ft.				Horse power

	A	Under head in ft.				
or,	B C	Set 1	R to head in ft.	158 to R	R to vent in sq. in.	1 to R	Under useful effect
	D						Horse power.

36. — Horse Power of a Steam Engine

C	Set 21,000	R to diam.	1 to R	R to stroke in ft.	1 to R	R to rev. per min.	1 to R	Mean pressure per sq. inch
D	To diam. in inches							Horse power

	A						Horse power
or,	B C	Set 21,000	R to stroke in ft.	1 to R	R to revolutions	1 to R	Mean pressure
	D	To diam. in inches					

164 TYPICAL PROBLEMS AND SLIDE RULE SETTINGS

37. — Dynamometer; to Estimate the H.P. Indicated by

C	Set 5252	R to length of lever in feet from center of shaft	1 to R	Under rev. of shaft per min.
D	Weight applied at end of lever in pounds, including weight of scale			Actual horse power.

C. TABLES OF CONVERSIONS AND GAUGE POINTS FOR C AND D SCALES.

The settings for the following conversions are made in the usual manner, as explained above in Section B. Since all are intended for the same scales, however (C and D), the scales are not indicated in the settings. The numbers and other quantities above the horizontal line in each case are to be read on the C scale and those below the line on the D scale. After the slide is set as indicated by the two numbers given for each setting the runner may be set at the quantity on the C scale or that on D, that is, the setting or conversion may be used in either sense. Thus as an example consider the first setting under GEOMETRICAL below, and let it be required to find the circumference of a circle of 12 inches diameter. With 226 on C set at 710 on D the runner is set at 12 on C and on D is read the circumference 37.7 inches. On the other hand, without moving the slide the same ratio may be used to find the diameter of a circle whose circumference is known to be 56.5 inches; thus set the runner at 56.5 on D and on C read the diameter 18.01 inches. Similarly the first setting under METRIC

TABLES OF CONVERSIONS AND GAUGE POINTS 165

SYSTEM may be used to convert inches to centimeters or centimeters to inches, and so for all the others. As further illustration we give here a few examples involving some of the COMBINATIONS.

EXAMPLES

What is the pressure in pounds per square inch equivalent to a head of 34 feet of water?

C	Set 60	Under 34
D	To 26	Find 14.75 pounds — Answer

What head of water, in feet, is equivalent to a pressure of 18 pounds per square inch?

C	Set 26	Under 18
D	To 60	Find 41.5 feet — Answer

How many horse power will 50 cubic feet of water per minute give under a head of 400 feet?

C	Set 3700	Runner to 400	1 to R	Under 50
D	To 7			Find 37.8 H.P. — Answer

In each of the equivalent ratios given in this section the error is less than one-tenth of one percent.

(1) GEOMETRICAL

Set 226	= Diameters of circles.
To 710	= Circumferences of circles.
79	= Diameter of circle.
70	= Side of equal square.
99	= Diameter of circle.
70	= Side of inscribed square.

$\dfrac{39}{11}$ = Circumference of circle.
  = Side of equal square.

$\dfrac{40}{9}$ = Circumference of circle.
 = Side of inscribed square.

$\dfrac{70}{99}$ = Side of square.
 = Diagonal of square.

$\dfrac{205}{161}$ = Area of square whose side = 1.
 = Area of circle whose diameter = 1.

$\dfrac{322}{205}$ = Area of circle.
 = Area of inscribed square.

(2) Arithmetical

$\dfrac{100}{66}$ = Links
 = Feet

$\dfrac{12}{95}$ = Links.
 = Inches.

$\dfrac{101}{44}$ = Square links.
 = Square feet.

$\dfrac{6}{5}$ = U. S. gallons.
 = Imperial gallons.

$\dfrac{1}{231}$ = U. S. gallons.
 = Cubic inches.

$\dfrac{800}{107}$ = U. S. gallons.
 = Cubic feet.

$\dfrac{22}{6100}$ = Imperial gallons.
 = Cubic inches.

$\dfrac{430}{69}$ = Imperial gallons.
 = Cubic feet.

TABLES OF CONVERSIONS AND GAUGE POINTS

(3) METRIC SYSTEM

$$\frac{26 = \text{Inches.}}{66 = \text{Centimetres.}}$$

$$\frac{82 = \text{Yards.}}{75 = \text{Metres.}}$$

$$\frac{4300 = \text{Links.}}{865 = \text{Metres.}}$$

$$\frac{82 = \text{Feet.}}{25 = \text{Metres.}}$$

$$\frac{87 = \text{Miles.}}{140 = \text{Kilometres.}}$$

$$\frac{43 = \text{Chains.}}{865 = \text{Metres.}}$$

$$\frac{31 = \text{Square inches.}}{200 = \text{Square centimetres.}}$$

$$\frac{140 = \text{Square feet.}}{13 = \text{Square metres.}}$$

$$\frac{61 = \text{Square yards.}}{51 = \text{Square metres.}}$$

$$\frac{42 = \text{Acres.}}{17 = \text{Hectares.}}$$

$$\frac{22 = \text{Square miles.}}{57 = \text{Square kilometres.}}$$

$$\frac{5 = \text{Cubic inches.}}{82 = \text{Cubic centimetres.}}$$

$$\frac{600 = \text{Cubic feet.}}{17 = \text{Cubic metres.}}$$

$$\frac{85 = \text{Cubic yards.}}{65 = \text{Cubic metres.}}$$

$$\frac{6 = \text{Cubic feet.}}{170 = \text{Litres.}}$$

$$\frac{14 = \text{U. S. gallons.}}{53 = \text{Litres.}}$$

$$\frac{46 = \text{Imperial gallons.}}{209 = \text{Litres.}}$$

168   TYPICAL PROBLEMS AND SLIDE RULE SETTINGS

$$\frac{108 = \text{Grains.}}{7 = \text{Grammes.}}$$

$$\frac{75 = \text{Pounds.}}{34 = \text{Kilogrammes.}}$$

$$\frac{6 = \text{Ounces.}}{170 = \text{Grammes.}}$$

$$\frac{63 = \text{Hundredweights.}}{3200 = \text{Kilogrammes.}}$$

$$\frac{63 = \text{English Tons.}}{64 = \text{Metric Tonnes.}}$$

(4) PRESSURES

$$\frac{640 = \text{Pounds per square inch.}}{45 = \text{Kilogs per square centimetre.}}$$

$$\frac{51 = \text{Pounds per square foot.}}{249 = \text{Kilogs per square metre.}}$$

$$\frac{59 = \text{Pounds per square yard.}}{32 = \text{Kilogs per square metre.}}$$

$$\frac{57 = \text{Inches of mercury.}}{28 = \text{Pounds per square inch.}}$$

$$82 = \text{Inches of mercury.}$$
$$5800 = \text{Pounds per square foot.}$$
$$720 = \text{Inches of water.}$$

$$\frac{26 = \text{Pounds per square inch.}}{74 = \text{Inches of water.}}$$

$$\frac{385 = \text{Pounds per square foot.}}{60 = \text{Feet of water.}}$$

$$\frac{26 = \text{Pounds per square inch.}}{5 = \text{Feet of water.}}$$

$$\frac{312 = \text{Pounds per square foot.}}{15 = \text{Inches of mercury.}}$$

$$17 = \text{Feet of water.}$$

TABLES OF CONVERSIONS AND GAUGE POINTS 169

$$\frac{99}{2960} = \frac{\text{Atmospheres.}}{\text{Inches of mercury.}}$$

$$\frac{34}{500} = \frac{\text{Atmospheres.}}{\text{Pounds per square inch.}}$$

$$\frac{34}{7200} = \frac{\text{Atmospheres.}}{\text{Pounds per square foot.}}$$

$$\frac{30}{31} = \frac{\text{Atmospheres.}}{\text{Kilogs per square centimetre.}}$$

$$\frac{23}{780} = \frac{\text{Atmospheres.}}{\text{Feet of water.}}$$

$$\frac{3}{31} = \frac{\text{Atmospheres.}}{\text{Metres of water.}}$$

$$\frac{29}{67} = \frac{\text{Pounds per square inch.}}{\text{Feet of water.}}$$

$$\frac{1}{10} = \frac{\text{Kilogs per square centimetre.}}{\text{Metres of water.}}$$

(5) COMBINATIONS

$$\frac{43}{64} = \frac{\text{Pounds per foot.}}{\text{Kilogs per metre.}}$$

$$\frac{127}{63} = \frac{\text{Pounds per yard.}}{\text{Kilogs per metre.}}$$

$$\frac{46}{25} = \frac{\text{Pounds per square yard.}}{\text{Kilogs per square metre.}}$$

$$\frac{49}{785} = \frac{\text{Pounds per cubic foot.}}{\text{Kilogs per cubic metre.}}$$

170 TYPICAL PROBLEMS AND SLIDE RULE SETTINGS

$$\frac{27}{16} = \frac{\text{Pounds per cubic yard.}}{\text{Kilogs per cubic metre.}}$$

$$\frac{89}{42} = \frac{\text{Cubic feet per minute.}}{\text{Litres per second.}}$$

$$\frac{700}{53} = \frac{\text{Imperial gallons per minute.}}{\text{Litres per second.}}$$

$$\frac{840}{53} = \frac{\text{U. S. gallons per minute.}}{\text{Litres per second.}}$$

$$\frac{38}{39} = \frac{\text{Weight of fresh water.}}{\text{Weight of sea water.}}$$

$$\frac{5}{312} = \frac{\text{Cubic feet of water.}}{\text{Weight in pounds.}}$$

$$\frac{1}{10} = \frac{\text{Imperial gallons of water.}}{\text{Weight in pounds.}}$$

$$\frac{3}{25} = \frac{\text{U. S. gallons of water.}}{\text{Weight in pounds.}}$$

$$\frac{50}{6} = \frac{\text{Pounds per U. S. gallon.}}{\text{Kilogs per litre.}}$$

$$\frac{10}{1} = \frac{\text{Pounds per imperial gallon.}}{\text{Kilogs per litre.}}$$

$$\frac{30}{25} = \frac{\text{Pounds per U. S. gallon.}}{\text{Pounds per imperial gallon.}}$$

$$\frac{3}{85} = \frac{\text{Cubic feet of water.}}{\text{Weight in kilogs.}}$$

$$\frac{46}{209} = \frac{\text{Imperial gallons of water.}}{\text{Weight in kilogs.}}$$

SETTINGS FOR THE POLYPHASE SLIDE RULE 171

$\dfrac{14 = \text{U. S. gallons of water.}}{53 = \text{Weight in kilogs.}}$

$\dfrac{44 = \text{Feet per second.}}{30 = \text{Miles per hour.}}$

$\dfrac{88 = \text{Yards per minute.}}{3 = \text{Miles per hour.}}$

$\dfrac{41 = \text{Feet per second.}}{750 = \text{Metres per minute.}}$

$\dfrac{82 = \text{Feet per minute.}}{25 = \text{Metres per minute.}}$

$\dfrac{340 = \text{Footpounds.}}{47 = \text{Kilogrammetres.}}$

$\dfrac{72 = \text{British horse power.}}{73 = \text{French horse power.}}$

$\dfrac{3700 = \text{One cubic foot of water per minute under one foot of head.}}{7 = \text{British horse power.}}$

$\dfrac{75 = \text{One litre of water per second under one metre of head.}}{1 = \text{French horse power.}}$

D. SETTINGS FOR THE POLYPHASE MANNHEIM
SLIDE RULE.

In the list below are ninety-four settings which are specially adapted to the Polyphase Mannheim slide rule. (This rule is also manufactured under such names as Maniphase, Multiplex, etc.) The settings are not

172 TYPICAL PROBLEMS AND SLIDE RULE SETTINGS

diagrammed as in Section B above but are very clearly and concisely stated so that there is no difficulty in performing the operations. These descriptions are merely concise forms of what we have in Chapters II and III called the "statement" of a setting or operation, and although the various steps in each statement are not numbered they are to be carried out in the order given in each case. Although these settings are especially adapted to slide rules carrying the inverted (CI) and cube (K) scales, those involving only the standard scales will of course also apply to the standard Mannheim rule.

Although the present list is fairly long it is not to be understood that it is intended as complete; the settings given are the simpler and more commonly useful ones but there are of course scores of others of which those given are merely illustrative or typical. By a study of those in the list the manner of working out the combination of scales for almost any operation will become evident or be suggested.

In Articles 19 and 26 methods of finding powers and roots higher than the third are given in connection with the A, B and K scales; these methods are used here. Thus the fourth root of a number is obtained by finding the square root of the square root. The sixth root is obtained by finding the square root of the cube root. The eighth root is the square root of the fourth root, etc.

SETTINGS FOR THE POLYPHASE SLIDE RULE 173

Expressions Which May be Read Directly by Means of the Indicator, without Setting the Slide

1. $x = a^2$, set Indicator to a on D, read x on A.
2. $x = a^3$, set Indicator to a on D, read x on K.
3. $x = \sqrt{a}$, set Indicator to a on A, read x on D.
4. $x = \sqrt[3]{a}$, set Indicator to a on K, read x on D.
5. $x = \sqrt{a^3}$, set Indicator to a on A, read x on K.
6. $x = \sqrt[3]{a^2}$, set Indicator to a on K, read x on A.
7. $x = \dfrac{1}{a}$, set Indicator to a on CI, read x on C.
8. $x = \dfrac{1}{a^2}$, set Indicator to a on CI, read x on B.
9. $x = \dfrac{1}{\sqrt{a}}$, set Indicator to a on B, read x on CI.

With Indices in Alignment

10. $x = \dfrac{1}{a^3}$, set Indicator to a on CI, read x on K.
11. $x = \dfrac{1}{\sqrt[3]{a}}$, set Indicator to a on K, read x on CI.

Expressions Solved with One Setting of Slide

SETTINGS FOR ONE FACTOR

12. $x = a^4$, set 1 to a on D, over a on C, read x on A.
13. $x = \dfrac{1}{a^4}$, set a on CI to a on D, under 1 on A, read x on B.
14. $x = a^5$, set a on CI to a on D, over a on B, read x on A.
15. $x = \dfrac{1}{a^5}$, set a on C to a on K, under a on CI, read x on K.
16. $x = a^6$, set 1 on C to a on D, under a on C, read x on K.
17. $x = a^7$, set a on CI to a on K, under a on C, read x on K.
18. $x = a^9$, set a on CI to a on D, under a on C, read x on K.
19. $x = \sqrt{a^5}$, set 1 to a on K, under a on B, read x on K.
20. $x = \sqrt{a^9}$, set 1 to a on A, under a on C, read x on K.
21. $x = \sqrt{a^{11}}$, set a on CI to a on K, under a on B, read x on K.
22. $x = \sqrt{a^{15}}$, set a on CI to a on D, under a on B, read x on K.

174   TYPICAL PROBLEMS AND SLIDE RULE SETTINGS

23. $x = \dfrac{1}{\sqrt[3]{a^2}}$, set 1 to a on K, under 1 on A, read x on B.

24. $x = \sqrt[3]{a^4}$, set 1 to a on K, under a on C, read x on D.

25. $x = \dfrac{1}{\sqrt[3]{a^4}}$, set a on CI to a on K, over 1 on D, read x on C.

26. $x = \sqrt[3]{a^5}$, set 1 to a on K, over a on B, read x on A.

27. $x = \dfrac{1}{\sqrt[3]{a^5}}$, set a on C to a on K, under a on CI, read x on D.

28. $x = \sqrt[3]{a^7}$, set a on CI to a on K, under a on C, read x on D.

29. $x = \sqrt[3]{a^8}$, set 1 to a on K, over a on C, read x on A.

30. $x = \dfrac{1}{\sqrt[3]{a^8}}$, set a on CI to a on K, under 1 on A, read x on B.

31. $x = \dfrac{1}{\sqrt[3]{a^{10}}}$, set a on C to a on K, over a on CI, read x on A.

32. $x = \sqrt[3]{a^{11}}$, set a on CI to a on K, over a on B, read x on A.

33. $x = \sqrt[3]{a^{14}}$, set a on CI to a on K, over a on C, read x on A.

34. $x = \sqrt[6]{a}$, set a on B to a on K, over 1 on D, read x on C.

35. $x = \dfrac{1}{\sqrt[6]{a}}$, set a on B to a on K, under 1 on C, read x on D.

36. $x = \sqrt[6]{a^5}$, set 1 to a on K, under a on B, read x on D.

37. $x = \sqrt[6]{a^7}$, set a on B to a on K, over a on D, read x on C.

38. $x = \dfrac{1}{\sqrt[6]{a^7}}$, set a on B to a on K, over a on D, read x on CI.

39. $x = \sqrt[6]{a^{11}}$, set a on CI to a on K, under a on B, read x on D

SETTINGS FOR TWO FACTORS

40. $x = ab$, set 1 to a on D, under b on C, read x on D.

41. $x = \dfrac{1}{ab}$, set a on CI to b on D, over 1 on D, read x on C

42. $x = \dfrac{a}{b}$, set b on C to a on D, under 1 on C, read x on D.

43. $x = \dfrac{1}{\frac{a}{b}}$, set b on C to a on D, over 1 on D, read x on C.

SETTINGS FOR THE POLYPHASE SLIDE RULE 175

44. $x = ab^2$, set 1 to a on A, over b on C, read x on A.
45. $x = \dfrac{1}{ab^2}$, set b on CI to a on A, under 1 on A, read x on B.
46. $x = \dfrac{a}{b^2}$, set b on C to a on A, over 1 on B, read x on A.
47. $x = \dfrac{a^2}{b}$, set a on C to b on A, under 1 on A, read x on B.
48. $x = a^2b^2$, set 1 on C to a on D, over b on C, read x on A.
49. $x = \dfrac{1}{a^2b^2}$, set a on CI to b on D, under 1 on A, read x on B.
50. $x = \dfrac{1}{a^3b^3}$, set a on C to 1 on D, under b on CI, read x on K.
51. $x = \dfrac{a^2}{b^2}$, set b on C to a on D, at 1 on C, read x on A.
52. $x = a\sqrt{b}$, set 1 to a on D, under b on B, read x on D.
53. $x = \dfrac{1}{a\sqrt{b}}$, set a on CI to b on A, over 1 on D, read x on C.
54. $x = \dfrac{\sqrt{a}}{b}$, set b on C to a on A, under 1 on C, read x on D.
55. $x = \dfrac{a}{\sqrt{b}}$, set b on B to a on D, under 1 on C, read x on D.
56. $x = a^2\sqrt{b}$, set a on CI to a on D, under b on B, read x on D.
57. $x = a^4b$, set a on CI to b on A, over a on C, read x on A.
58. $x = a^6\sqrt{b^3}$, set a on CI to b on A, under a on C, read x on K.
59. $x = \dfrac{a^2}{\sqrt{b}}$, set b on B to a on D, under a on C, read x on D.
60. $x = \dfrac{\sqrt{a}}{b^2}$, set b on C to a on A, under b on CI, read x on D.
61. $x = ab^3$, set 1 on C to a on K, under b on C, read x on K.
62. $x = \dfrac{a}{b^3}$, set b on C to a on K, under 1 on C, read x on K.
63. $x = \dfrac{a^3}{b^2}$, set a on CI to a on A, over b on CI, read x on A.
64. $x = \dfrac{a^2}{b^3}$, set b on B to a on D, over b on CI, read x on A.
65. $x = a^2b^2$, set b on CI to a on D, over b on B, read x on A.
66. $x = a\sqrt[3]{b}$, set 1 on C to b on K, under a on C, read x on D.

176 TYPICAL PROBLEMS AND SLIDE RULE SETTINGS

67. $1 \div a\sqrt[3]{b}$, set a on CI to b on K, over 1 on D, read x on C.
68. $x = a \div \sqrt[3]{b}$, set a on C to b on K, over 1 on D, read x on C.
69. $x = \sqrt[3]{a} \div b$, set b on C to a on K, under 1 on C, read x on D.
70. $x = a^3 b^3$, set 1 on C to b on D, under a on C, read x on K.
71. $x = \dfrac{a^3}{b^3}$, set b on C to a on D, under 1 on C read x on K.
72. $x = ab^4$, set b on CI to a on A, over b on C, read x on A.
73. $x = b^6\sqrt{a^3}$, set b on CI to a on A, under b on C, read x on K.
74. $x = a^2\sqrt[3]{b^2}$, set 1 on C to b on K, over a on C, read x on A.
75. $x = \dfrac{\sqrt[3]{a^4}}{b}$, set b on C to a on K, under a on C, read x on D.
76. $x = \dfrac{\sqrt[3]{a^8}}{b^2}$, set b on C to a on K, over a on C, read x on A.
77. $x = \dfrac{a^4}{b^3}$, set b on C to a on K, over a on C, read x on K.

SETTINGS FOR THREE FACTORS

78. $x = a \cdot b \cdot c$, set a on CI to b on D, under c on C, read x on D
79. $x = a^2 \times b^2 \times c^2$, set a on CI to b on D, over c on C, read x on A.
80. $x = a^3 \times b^3 \times c^3$, set a on CI to b on D, over c on C, read x on K.
81. $x = \dfrac{a \times b}{c}$, set c on C to a on D, under b on C, read x on D.
82. $x = \dfrac{a^2 b^2}{c^2}$, set c on C to a on D, over b on C, read x on A.
83. $x = \dfrac{a^3 b^3}{c^3}$, set c on C to a on D, over b on C, read x on K.
84. $x = \dfrac{a}{b \cdot c}$, set b on C to a on D, under c on CI, read x on D.
85. $x = \dfrac{a^2}{b^2 \times c^2}$, set b on C to a on D, over c on CI, read x on A.
86. $x = \dfrac{a^3}{b^3 \times c^3}$, set b on C to a on D, over c on CI, read x on K.
87. $x = ab\sqrt{c}$, set a on CI to c on A, under b on C, read x on D.
88. $x = a^2 b^2 c$, set a on CI to c on A, over b on C, read x on A.
89. $x = a^3 b^3 \sqrt{c^3}$, set a on CI to c on A, over b on C, read x on K.
90. $x = ab\sqrt[3]{c}$, set a on CI to c on K, under b on C, read x on D.

SETTINGS FOR POLYPHASE DUPLEX AND LOG-LOG 177

91. $x = a^2 b^2 \sqrt[3]{c^2}$, set a on CI to c on K, over b on C, read x on A.
92. $x = a^3 \cdot b^3 \cdot c$, set a on CI to c on K, over b on C, read x on K.
93. $x = \dfrac{\sqrt{a}}{b\sqrt[3]{c}}$, set b on CI to c on K, under a on A, read x on C.
94. $x = \dfrac{a\sqrt{b}}{\sqrt[3]{c}}$, set a on C, to c on K, under b on A, read x on C.

And scores of other combinations.

E. SETTINGS FOR THE POLYPHASE DUPLEX AND LOG-LOG DUPLEX SLIDE RULES.

In the following list of 153 settings the method of writing the formulas and stating the settings is the same as that used above in Section D. Since the Log-Log Duplex rule carries all the scales found on the Polyphase Duplex these settings are equally well adapted to either rule. It is to be understood in using each of these rules that when a setting of slide or runner (indicator) or both has been made it may be necessary to turn the rule over in order to read the result which may be on the face opposite from the one on which the first steps in the setting were carried out; also it may be necessary to turn the rule over several times in the course of a complete operation of several steps. It is also to be understood as in Section D that this list is by no means exhaustive.

Expressions Read Directly by Means of the Indicator without Setting the Slide

1. $x = a^2$, set Indicator to a on D, read x on A.
2. $x = a^3$, set Indicator to a on D, read x on K.
3. $x = \sqrt{a}$, set Indicator to a on A, read x on D.

178 TYPICAL PROBLEMS AND SLIDE RULE SETTINGS

4. $x = \sqrt[3]{a}$, set Indicator to a on K, read x on D.
5. $x = \sqrt{a^3}$, set Indicator to a on A, read x on K.
6. $x = \sqrt[3]{a^2}$, set Indicator to a on K, read x on A.
7. $x = \dfrac{1}{a}$, set Indicator to a on CI, read x on C.
8. $x = a \times \pi$, set Indicator to a on D, read x on DF.
9. $x = \dfrac{a}{\pi}$, set Indicator to a on DF, read x on D.
10. $x = \dfrac{\pi}{a}$, set Indicator to a on CF, read x on CI.
11. $x = \dfrac{1}{a \times \pi}$, set Indicator to a on C, read x on CIF.
12. $x = \pi\sqrt{a}$, set Indicator to a on A, read x on DF.
13. $x = \pi\sqrt[3]{a}$, set Indicator to a on K, read x on DF.
14. $x = \dfrac{a^2}{\pi^2}$, set Indicator to a on DF, read x on A.
15. $x = \dfrac{a^3}{\pi^3}$, set Indicator to a on DF, read x on K.
16. $x = \pi\sqrt{\sin a}$, set Indicator to a on S, read x on CF.
17. $x = \dfrac{1}{\pi\sqrt{\sin a}}$, set Indicator to a on S, read x on CIF.
18. $x = \pi \tan a$, set Indicator to a on T, read x on CF.
19. $x = \dfrac{1}{\pi \tan a}, = \dfrac{\cot a}{\pi}$, set Indicator to a on T, read x on CIF
20. $x = \dfrac{1}{\tan a} = \cot a$, set Indicator to a on T, read x on CI.
21. $x = \log a$, set Indicator to a on D, read x on L.
22. $x = \log\sqrt{a}$, set Indicator to a on A, read x on L.
23. $x = \log\sqrt[3]{a}$, set Indicator to a on K, read x on L.
24. $x = \log \dfrac{a}{\pi}$, set Indicator to a on DF, read x on L.
25. $x = \dfrac{1}{a^2}$, set Indicator to a on CI, read x on B.
26. $x = \dfrac{1}{\sqrt{a}}$, set Indicator to a on B, read x on CI.

SETTINGS FOR POLYPHASE DUPLEX AND LOG-LOG 179

With indices in alignment

27. $x = \tan a$, set Indicator to a on T, read x on D.
28. $x = \dfrac{1}{a^3}$, set Indicator to a on CI, read x on K.
29. $x = \dfrac{1}{\pi^2 a^2}$, set Indicator to a on CIF, read x on A.
30. $x = \dfrac{1}{\pi^3 a^3}$, set Indicator to a on CIF, read x on K.
31. $x = \dfrac{1}{\sqrt[3]{a}}$, set Indicator to a on K, read x on CI.
32. $x = \dfrac{1}{\pi\sqrt{a}}$, set Indicator to a on A, read x on CIF.
33. $x = \dfrac{1}{\pi\sqrt[3]{a}}$, set Indicator to a on K, read x on CIF.

Expressions Solved at One Setting of Slide
SETTINGS FOR ONE FACTOR

34. $x = a^4$, set 1 to a on D, at a on C, read x on A.
35. $x = \dfrac{1}{a^4}$, set a on CI to a on D, under 1 on A, read x on B.
36. $x = a^5$, set a on CI to a on D, over a on B, read x on A.
37. $x = \dfrac{1}{a^5}$, set a on C to a on K, at a on CI, read x on K.
38. $x = a^6$, set 1 on C to a on D, at a on C, read x on K.
39. $x = a^7$, set a on CI to a on K, at a on C, read x on K.
40. $x = a^9$, set a on CI to a on D, at a on C, read x on K.
41. $x = \sqrt{a^5}$, set 1 to a on K, at a on B, read x on K.
42. $x = \sqrt{a^9}$, set 1 to a on A, at a on C, read x on K.
43. $x = \sqrt{a^{11}}$, set a on CI to a on K, at a on B, read x on K.
44. $x = \sqrt{a^{15}}$, set a on CI to a on D, at a on B, read x on K.
45. $x = \dfrac{1}{\sqrt[3]{a^2}}$, set 1 to a on K, at 1 on A, read x on B.
46. $x = \sqrt[3]{a^4}$, set 1 to a on K, at a on C, read x on D.
47. $x = \dfrac{1}{\sqrt[3]{a^4}}$, set a on CI to a on K, at 1 on D, read x on C.

180   TYPICAL PROBLEMS AND SLIDE RULE SETTINGS

48. $x = \sqrt[3]{a^5}$, set 1 to a on K, at a on B, read x on A.
49. $x = \dfrac{1}{\sqrt[3]{a^5}}$, set a on C to a on K, at a on CI, read x on D.
50. $x = \sqrt[3]{a^7}$, set a on CI to a on K, at a on C, read x on D.
51. $x = \sqrt[3]{a^8}$, set 1 to a on K, at a on C, read x on A.
52. $x = \dfrac{1}{\sqrt[3]{a^8}}$, set a on CI to a on K, at 1 on A, read x on B.
53. $x = \dfrac{1}{\sqrt[3]{a^{10}}}$, set a on C to a on K, at a on CI, read x on A.
54. $x = \sqrt[3]{a^{11}}$, set a on CI to a on K, over a on C, read x on A.
55. $x = \sqrt[3]{a^{14}}$, set a on CI to a on K, at a on B, read x on A.
56. $x = \sqrt[6]{a}$, set a on B to a on K, at 1 on D, read x on C.
57. $x = \dfrac{1}{\sqrt[6]{a}}$, set a on B to a on K, at 1 on C, read x on D.
58. $x = \sqrt[6]{a^5}$, set 1 to a on K, at a on B, read x on D.
59. $x = \sqrt[6]{a^7}$, set a on B to a on K, at a on D, read x on C.
60. $x = \dfrac{1}{\sqrt[6]{a^7}}$, set a on B to a on K, at a on D, read x on CI.
61. $x = \sqrt[6]{a^{11}}$, set a on CI to a on K, at a on B, read x on D.

SETTINGS FOR TWO FACTORS

62. $x = ab$, set 1 to a on D, under b on C, read x on D.
63. $x = \dfrac{1}{ab}$, set a on CI to b on D, over 1 on D, read x on C.
64. $x = \dfrac{a}{b}$, set b on C to a on D, under 1 on C, read x on D.
65. $x = \dfrac{1}{\frac{a}{b}}$, set b on C to a on D, over 1 on D, read x on C.
66. $x = ab^2$, set 1 to a on A, over b on C, read x on A.
67. $x = \dfrac{1}{ab^2}$, set b on CI to a on A, under 1 on A, read x on B.
68. $x = \dfrac{a}{b^2}$, set b on C to a on A, over 1 on B, read x on A.

SETTINGS FOR POLYPHASE DUPLEX AND LOG-LOG 181

69. $x = \dfrac{a^2}{b}$, set a on C to b on A, under 1 on A, read x on B.

70. $x = a^2b^2$, set 1 on C to a on D, over b on C, read x on A.

71. $x = \dfrac{a^2}{a^2b^2}$, set a on C to 1 on D, over b on CI, read x on A.

72. $x = \dfrac{1}{a^3b^3}$, set a on C to 1 on D, over b on CI, read x on K.

73. $x = \dfrac{a^2}{b^2}$, set b on C to a on D, over 1 on C, read x on A.

74. $x = a\sqrt{a}$, set 1 on C to b on A, under a on C, read x on D.

75. $x = \dfrac{1}{a\sqrt{b}}$, set a on CI to b on A, over D, read x on C.

76. $x = \dfrac{\sqrt{a}}{b}$, set b on C to a on A, under 1 on C, read x on D.

77. $x = \dfrac{a}{\sqrt{b}}$, set b on B to a on D, under 1 on CI, read x on D.

78. $x = a^2\sqrt{b}$, set a on CI to a on D, under b on B, read x on D.
79. $x = a^4b$, set a on CI to b on A, over a on C, read x on A.
80. $x = a^6\sqrt{b^3}$ set a on CI to b on A, over 1 on C, read x on K.

81. $x = \dfrac{a^2}{\sqrt{b}}$, set b on B to a on D, over a on C, read x on D.

82. $x = \dfrac{\sqrt{a}}{b^2}$, set b on C to a on A, under b on CI, read x on D.

83. $x = ab^3$, set 1 on C to a on K, at b on C, read x on K.

84. $x = \dfrac{a}{b^3}$, set b on C to a on K, at 1 on C, read x on K.

85. $x = \dfrac{a^3}{b^2}$, set a on CI to a on A, over b on CI, read x on A.

86. $x = \dfrac{a^2}{b^3}$, set b on B to a on D, over b on CI, read x on A.

87. $x = a^2b^3$, set b on CI to a on D, over b on B, read x on A.

88. $x = \dfrac{a}{\sqrt[3]{b}}$, set 1 on C to b on K, over a on D, read x on A.

89. $x = \dfrac{1}{a\sqrt[3]{b}}$, set a on CI to b on K, over 1 on D, read x on C.

90. $x = \dfrac{a}{\sqrt[3]{b}}$, set a on C to b on K, over 1 on D, read x on C.

182 TYPICAL PROBLEMS AND SLIDE RULE SETTINGS

91. $x = \dfrac{\sqrt[3]{a}}{b}$, set b on C to a on K, under 1 on C, read x on D.

92. $x = a^3b^3$, set 1 on C to b on D, at a on C, read x on K.

93. $x = \dfrac{a^3}{b^3}$, set b on C to a on D, at 1 on C, read x on K.

94. $x = ab^4$, set b on CI to a on A, over b on C, read x on A.

95. $x = \sqrt[6]{a^3}$, set b on CI to a on A, at b on C, read x on K.

96. $x = a^2\sqrt[3]{b^2}$, set 1 on C to b on K, under a on C, read x on A.

97. $x = \dfrac{\sqrt[3]{a^4}}{b}$, set b on C to a on K, under a on C, read x on D.

98. $x = \dfrac{\sqrt[3]{a^8}}{b^2}$, set b on C to a on K, over a on C, read x on A.

99. $x = \dfrac{a^4}{b^3}$, set b on C to a on K, over a on C, read x on K.

SETTINGS FOR TWO FACTORS AND π

100. $x = a \cdot b \cdot \pi$, set 1 on C to a on D, over b on C, read x on DF.

101. $x = \dfrac{\pi}{ab}$, set a on CI to b on D, over 1 on D, read x on CF.

102. $x = \dfrac{ab}{\pi}$, set a on CI to b on D, over 1 on D, read x on CIF.

103. $x = \dfrac{\pi^2 a}{b^2}$, set b on CF to a on A, over 1 on C, read x on A.

104. $x = \dfrac{\pi\sqrt{a}}{b}$, set b on CF to a on A, under 1 on C, read x on D.

105. $x = \dfrac{\pi^2}{a^2b^2}$, set a on CF to 1 on D, over b on CI, read x on A.

106. $x = \dfrac{\pi^3}{a^3b^3}$, set a on CF to 1 on D, over b on CI, read x on K.

107. $x = \dfrac{\pi^2 a^2}{b^2}$, set b on CF to 1 on D, over a on C, read x on A.

108. $x = \dfrac{\pi^3 a^3}{b^3}$, set b on CF to 1 on D, over a on C, read x on K.

109. $x = \dfrac{a}{\pi b}$, set 1 on C to a on D, under b on CIF, read x on D.

110. $x = \dfrac{a^2}{\pi^2 b^2}$, set 1 on C to a on D, over b on CIF, read x on K.

SETTINGS FOR POLYPHASE DUPLEX AND LOG-LOG 183

111. $x = \dfrac{a^3}{\pi^3 b_2}$, set 1 on C to a on D, over b on CIF, read x on K.

112. $x = \pi a^2 \sqrt{b}$, set a on CI to b on A, over a on C, read x on DF.

113. $x = \dfrac{\pi a}{\sqrt{b}}$, set a on C to b on A, over 1 on D, read x on CF.

114. $x = \dfrac{\sqrt{b}}{\pi a}$, set a on C to b on A, over 1 on D, read x on CIF.

115. $x = \dfrac{\pi a^2}{\sqrt{b}}$, set a on C to b on A, over a on D, read x on CF.

116. $x = \dfrac{\sqrt{b}}{\pi a^2}$, set a on C to b on A, over a on D, read x on CIF.

117. $x = \dfrac{\pi \sqrt[3]{a}}{b}$, set b on C to a on K, over 1 on C, read x on DF.

118. $x = \pi \sqrt{a^5}$, set a on CI to a on A, over a on C, read x on DF.

119. $x = \pi \sqrt[3]{a^4}$, set 1 to a on K, over a on C, read x on DF.

120. $x = \dfrac{\pi}{\sqrt[3]{a^4}}$, set a on CI to a on K, over 1 on D, read x on CF.

121. $x = \dfrac{\pi}{\sqrt[3]{a^5}}$, set a on C to a on K, over a on CI, read x on DF.

122. $x = \pi \sqrt[3]{a^7}$, set a on CI to a on K, over a on C, read x on DF.

123. $x = \dfrac{\pi}{\sqrt[3]{a^2}}$, set a on C to a on K, at 1 on C, read x on DF.

124. $x = \dfrac{\pi}{\sqrt[3]{a^5}}$, set a on C to a on K, at a on CI, read x on DF.

125. $x = \dfrac{\pi \sqrt[3]{a^4}}{b}$, set b on C to a on K, over a on C, read x on DF.

126. $x = \dfrac{\pi}{a \sqrt[3]{b}}$, set a on CI to b on K, over 1 on D, read x on CF.

127. $x = \dfrac{a \sqrt[3]{b}}{\pi}$, set a on CI to b on K, over 1 on D, read x on CIF.

SETTINGS FOR THREE FACTORS

128. $x = a \cdot b \cdot c$, set a on CI to b on D, under c on C, read x on D.
129. $x = a^2 \times b^2 \times c^2$, set a on CI to b on D, over c on C, read x on A.
130. $x = a^3 \times b^3 \times c^3$, set a on CI to b on D, over c on C, read x on K.

184 TYPICAL PROBLEMS AND SLIDE RULE SETTINGS

131. $x = \dfrac{a \times b}{c}$, set c on C to a on D, under b on C, read x on D.

132. $x = \dfrac{a^2 b^2}{c^2}$, set c on C to a on D, over b on C, read x on A.

133. $x = \dfrac{a^3 b^3}{c^3}$, set c on C to a on D, over b on C, read x on K.

134. $x = \dfrac{a}{b \cdot c}$, set b on C to a on D, under c on CI, read x on D.

135. $x = \dfrac{a^2}{b^2 \times c^2}$, set b on C to a on D, over c on CI, read x on A.

136. $x = \dfrac{a^3}{b^3 \times c^3}$, set b on a C to on D, over c on CI, read x on K.

137. $x = ab\sqrt{c}$, set a on CI to c on A, under b on C, read x on D.
138. $x = a^2 b^2 c$, set a on CI to c on A, over b on C, read x on A.
139. $x = a^3 b^3 \sqrt{c^3}$, set a on CI to c on A, over b on C, read x on K.
140. $x = ab\sqrt[3]{c}$, set a on CI to c on K, under b on C, read x on D.
141. $x = a^2 b^2 \sqrt[3]{c^2}$, set a on CI to c on K, over b on C, read x on A.
142. $x = a^3 \cdot b^3 \cdot c$, set a on CI to c on K, over b on C, read x on K.

143. $x = \dfrac{\sqrt{a}}{b\sqrt[3]{c}}$, set b on CI to c on K, under a on A, read x on C.

144. $x = \dfrac{a\sqrt{b}}{\sqrt[3]{c}}$, set a on C, to c on K, under b on A, read x on C.

And scores of other combinations.

SETTINGS FOR THREE FACTORS AND π

145. $a \cdot b \cdot c \cdot \pi$, set a on CI to b on D, over c on C, read x on DF.

146. $x = \dfrac{ab\pi}{c}$, set c on C to a on D, over b on C, read x on DF.

147. $x = \dfrac{a\pi}{bc}$, set b on C to a on D, over c on CI, read x on DF.

148. $x = ab\pi\sqrt{c}$, set a on CI to c on A, over b on C, read x on DF.
149. $x = ab\pi\sqrt[3]{c}$, set a on CI to c on K, over b on C, read x on DF.

150. $x = \dfrac{\pi\sqrt{a}}{b\sqrt[3]{c}}$, set b on CI to c on K, under a on A, read x on CF.

151. $x = \dfrac{b\sqrt[3]{c}}{\pi\sqrt{a}}$, set b on CI to c on K, under a on A, read x on CIF.

SETTINGS INVOLVING THE LOG-LOG SCALES 185

152. $x = \dfrac{\pi a \sqrt{b}}{\sqrt[3]{c}}$, set a on C to c on K, under b on A, read x on CF.

153. $x = \dfrac{\sqrt[3]{c}}{\pi a \sqrt{b}}$, set a on C to c on K, under b on A, read x on CIF.

F. SETTINGS AND TYPICAL PROBLEMS INVOLVING THE LOG-LOG SCALES.

In the last list above (Section E) are given settings which involve all the scales of the Log-log Duplex slide rule except the log-log scales. In the following list, therefore, are included only a few such settings, the greater part of the list consisting of settings which involve the log-log scales.

Following the list of sixty-one settings and numbered continuously with these are given the diagrams for twenty-eight typical formulas and problems from various branches of physics and engineering which involve higher powers and roots and exponentials and for which the log-log scales are particularly adapted. By studying these problems and their settings one can easily work out the settings for similar types of problems. For purposes of comparison with the types of settings given in Section B, the first three problems in this second part of the list are taken from Section B and are worked out both with and without the use of the log-log scales or by combination of the two methods.

186 TYPICAL PROBLEMS AND SLIDE RULE SETTINGS

Expressions which may be Read Directly by Means of the Indicator, without Setting the Slide

1. $x = a^2$, set Indicator to a on D, read x on A.
2. $x = a^3$, set Indicator to a on D, read x on K.
3. $x = a^{10}$, set Indicator to a on LL1, read x on LL2; or set Indicator to a on LL2, read x on LL3.
4. $x = a^{100}$, set Indicator to a on LL1, read x on LL3.
5. $x = \sqrt{a}$, set Indicator to a on A, read x on D.
6. $x = \sqrt[10]{a}$, set Indicator to a on LL3, read x on LL2; or set Indicator to a on LL2, read x on LL1.
7. $x = \sqrt[100]{a}$, set Indicator to a on LL3, read x on LL1.
8. $x = \sqrt[3]{a}$, set Indicator to a on K, read x on D.
9. $x = \sqrt{a^3}$, set Indicator to a on A, read x on K.
10. $x = \sqrt[3]{a^2}$, set Indicator to a on K, read x on A.
11. $x = \dfrac{1}{a}$, set Indicator to a on CI, read x on C.
12. $x = a \times \pi$, set Indicator to a on D, read x on DF.
13. $x = \dfrac{a}{\pi}$, set Indicator to a on DF, read x on D.
14. $x = \dfrac{\pi}{a}$, set Indicator to a on CF, read x on CI.
15. $x = \dfrac{1}{a \times \pi}$, set Indicator to a on C, read x on CIF.
16. $x = \pi\sqrt{a}$, set Indicator to a on A, read x on DF.
17. $x = \pi\sqrt[3]{a}$, set Indicator to a on K, read x on DF.
18. $x = \dfrac{a^2}{\pi^2}$, set Indicator to a on DF, read x on A.
19. $x = \dfrac{a^3}{\pi^3}$, set Indicator to a on DF, read x on K.
20. $x = \pi\sqrt{\sin a}$, set Indicator to a on S, read x on CF.
21. $x = \dfrac{1}{\pi\sqrt{\sin a}}$, set Indicator to a on S, read x on CIF.
22. $x = \pi \tan a$, set Indicator to a on T, read x on CF.
23. $x = \dfrac{1}{\pi \tan a} = \dfrac{\cot a}{\pi}$, set Indicator to a on T, read x on CIF.
24. $x = \tan a$, set Indicator to a on T, read x on C.

SETTINGS INVOLVING THE LOG-LOG SCALES

25. $x = \dfrac{1}{\tan a} = \cot a$, set Indicator to a on T, read x on CI.
26. $x = \log a$, set Indicator to a on D, read x on L.
27. $x = \log_e a$, set Indicator to a on LL1, LL2 or LL3, read x on D.
28. $x = \colog_e a$, set Indicator to a on LL0, read x on A.
29. $x = \log\sqrt{a}$, set Indicator to a on A, read x on L.
30. $x = \log\sqrt[3]{a}$, set Indicator to a on K, read x on L.
31. $x = \log\dfrac{a}{\pi}$, set Indicator to a on DF, read x on L.

With Indices in Alignment

32. $x = \dfrac{1}{a^2}$, set Indicator to a on CI, read x on A.
33. $x = \dfrac{1}{a^3}$, set Indicator to a on CI, read x on K.
34. $x = \dfrac{1}{\pi^2 a^2}$, set Indicator to a on CIF, read x on A.
35. $x = \dfrac{1}{\pi^3 a^3}$, set Indicator to a on CIF, read x on K.
36. $x = \dfrac{1}{\sqrt{a}}$, set Indicator to a on A, read x on CI.
37. $x = \dfrac{1}{\sqrt[3]{a}}$, set Indicator to a on K, read x on CI.
38. $x = \dfrac{1}{\pi\sqrt{a}}$, set Indicator to a on A, read x on CIF
39. $x = \dfrac{1}{\pi\sqrt[3]{a}}$, set Indicator to a on K, read x on CIF.

Expressions Solved at One Setting of Slide

40. $x = a^4$, set 1 to a on D, at a on C, read x on A; or set 1 to a on LL, at 4 on B or C, read x on LL.
41. $x = \dfrac{1}{a^5}$, set a on C to a on K, at a on CI, read x on K.
42. $x = a^6$, set 1 on C to a on D, at a on C, read x on K; or set 1 to a on LL, at 6 on B or C, read x on LL.

188 TYPICAL PROBLEMS AND SLIDE RULE SETTINGS

43. $x = a^7$, set a on CI to a on K, at a on C, read x on K; or set 1 to a on LL, at 7 on B or C, read x on LL.
44. $x = a^9$, set a on CI to a on D, at a on C, read x on K; or set 1 to a on LL, at 9 on B or C, read x on LL.
45. $x = a^n$, set 1 to a on LL, at n on B or C, read x on LL.
46. $x = \sqrt[n]{a}$, set n to a on LL, at 1 on B or C, read x on LL.
47. $x = \sqrt{a^9}$, set 1 to a on A, at a on C, read x on K.
48. $x = \sqrt[3]{a^4}$, set 1 to a on K, at a on C, read x on D.
49. $x = \dfrac{1}{\sqrt[3]{a^4}}$, set a on CI, to a on K, at 1 on D, read x on C.
50. $x = \dfrac{1}{\sqrt[3]{a^5}}$, set a on C to a on K, at a on CI, read x on D.
51. $x = \sqrt[3]{a^7}$, set a on CI to a on K, at a on C, read x on D.
52. $x = \sqrt[3]{a^8}$, set 1 to a on K, at a on C, read x on A.
53. $x = \dfrac{1}{\sqrt[3]{a^{10}}}$, set a on C to a on K, at a on CI, read x on A.
54. $x = \sqrt[3]{a^{14}}$, set a on CI to a on K, over a on C, read x on A.
55. $x = \dfrac{1}{ab}$, set a on CI to b on D, at 1 on D, read x on C.
56. $x = a^5$, set a on CI to a on A, over a on C, read x on A.
57. $x = \sqrt{a^5}$, set a on CI to a on A, at a on C, read x on D.
58. $x = \dfrac{1}{\sqrt[3]{a^2}}$, set a on C to a on K, at 1 on C, read x on D.
59. $x = \dfrac{1}{\sqrt[3]{a^5}}$, set a on C to a on K, at a on CI, read x on D.
60. $x = \log a$, set 1 on C to 10 on LL3, at a on LL1, LL2, or LL3, read x on C.
61. $x = \operatorname{colog} a$, set 1 to 10 on LL0, at a on LL0, read x on B.

62. **Strength of Wrought Iron Shafting:**

$$D = \sqrt[3]{\dfrac{83\,H}{N}} \text{ for crank shafts and prime movers}$$

$$D = \sqrt[3]{\dfrac{65\,H}{N}} \text{ for ordinary shafting}$$

SETTINGS INVOLVING THE LOG-LOG SCALES 189

C	Set R.P.M.	Indicator to I.H.P.	
D	To 83 or 65	Read diameter[3]	Read dia.
K			Opposite dia.[3]

			or
C		Set 3	Opposite index
D			
LL		To diameter[3]	Read diameter

NOTE. — In this, as in other cases, the coefficients (83 and 65) may be altered to suit individual opinions, without in any way altering the methods of solution.

63. Compound Interest:

When r is the interest rate expressed in hundredths of a dollar, n is the number of interest periods (years, half-years, quarters, etc.), P is the principal in dollars, and A is the amount in dollars, the formula is
$$A = P(1 + r)^n$$
therefore $\quad \text{Log } A = \text{Log } P + n \text{ Log}(1 + r)$

Opposite 1 plus the rate of interest on D, find the corresponding number on the L scale and multiply it by the number of years.

Set the indicator to this product on the L scale.

Set index of C to the indicator.

Opposite the principal on C, read the amount for the given number of years at the given rate on D

D	Set Ind. to 105			Read $244.35 Ans.
L	Read .0212	Ind. to .2(.0212×10)		
C			Left index to ind.	Opposite 150

This problem is solved simply on the *Log-Log scales* as follows:
$$A = P(1 + r)^n$$
$$1 + r = 1.05$$
$$n = 10$$

190 TYPICAL PROBLEMS AND SLIDE RULE SETTINGS

To 1.05 on LL1, set indicator.
Since the values on the LL2 scale are the 10th powers of the quantities on the LL1 scale, read 1.63 under the indicator on the LL2 scale.
$$1.63 = (1 + r)^n$$
To 1.63 on D, set left index of C.
Below 150 (P) on C, read 244.35 on D.
We thus obtain on D, below 1 on C, a gauge-point for 10 years at 5 per cent and can obtain in like manner similar ones for any other number of years and rate of interest.

64. **Delivery of Water from Pipes:** $W = 4.71 \sqrt{\dfrac{D^5 H}{L}}$

Eytelwein's Rule

L	Read log of dia.	Log dia $\times 5 = x$		
D	Opp. dia. in inch.			Read cu. ft. per min.
A		Opposite x		
B		Set length in feet	Opp. head in feet	
C			Set index	Opposite 4.71

When setting x on A do not include characteristic.

Or: —

LL	To dia. in in.	Read dia.[5]			
C	Set index	Below 5		Index to indicator	Opposite 4.71
A			Opposite dia.[5]		
B			Set length in feet	Indicator to head in feet	
D					Read cubic feet per minute

65. **Cylindrical Columns:**

For solid cylindrical columns of cast iron, both ends rounded, the length of the column exceeding 15 times the diameter
$$P = 33{,}380 \frac{d^{3.76}}{L^{1.7}}$$

SETTINGS INVOLVING THE LOG-LOG SCALES 191

where P = crushing weight in pounds; d = exterior diameter in inches; L = length in feet.

LL	To d	Read $d^{3.76}$	To L	Read $L^{1.7}$		
C	Set 1	Below 3.76	Set 1	Below 1.7	Set $L^{1.7}$	Below 33,380
D					To $d^{3.76}$	Read P

66. Resistance of Hollow Cylinders to Collapse:
Fairbairn's Empirical Formula

$$p = 9{,}675{,}000 \frac{t^{2.19}}{ld}$$

where p = pressure in lbs. per square inch; t = thickness of cylinder in inches; d = diameter in inches; l = length in inches.

LL	To t	Read $t^{6.15}$		
B or C	Set 1	Opposite 2.19		
D			To $t^{6.15}$	At Indicator read p
C			Set d	
CI			Indicator to l	Set 9,675,000 to Ind.

67. Torsional Strength:
For hollow shaft:

$$\text{Pa} = .1963 \frac{d^4 - d_1^4}{d} S$$

Pa = moment of the applied force.
d = external diameter of shaft.
d_1 = internal diameter of shaft.
S = unit shearing resistance.

LL	To d_1	Read d^4	To d_1	Read d_1^4			
C	Set 1	Opp. 4	Set 1	Opp. 4	Set d	Ind. to .1963	Opp. Ind.
D					To $d^4 - d_1^4$		Read Pa
CI					S to Ind.		

68. Radius of Gyration:
Spherical shell, radii R, r; revolving on its dia.

$$G = .6325 \sqrt{\frac{R^5 - r^5}{R^3 - r^3}}$$

192   TYPICAL PROBLEMS AND SLIDE RULE SETTINGS

LL	To R	Read R^5	Read R^3	To r	Read r^5	Read r^3		
C or B	Set 1	Opp. 5	Opp. 3	Set 1	Opp. 5	Opp. 3		Index to .6325
A							To R^5-r^5	
B							Set R^3-r^3	
D								Read G

69. Flow of Air in Pipes:

$$Q = 3.287 \sqrt{\frac{pd^5}{L}}$$

Q = Quantity in cubic feet per second.
p = Head or pressure in lbs. per square inch.
d = Diameter in inches.
L = Length in feet.

LL	To d	Read d^5					
C or B	Set 1	Opposite 5	C			1 to d	Below 3.287
A				To d^5			
			B	Set L	Ind. to p		
			D			Read Q	Read Q

70. Work of Adiabatic Compression of Air:

Mean effective pressure during the stroke $= 3.463\, p_1 \left\{ \left(\frac{p_2}{p_1}\right)^{0.29} - 1 \right\}$

Where p_1 and p_2 are absolute pressures above a vacuum in atmospheres or in pounds per square inch or per square foot.

Example: Required the work done in compressing one cubic foot of air per second from 1 to 6 atmospheres, including the work of expulsion from the cylinder.

$$\frac{p_2}{p_1} = 6$$

LL3	To 6					
C	Set 1	Opp. 29	Set 10	Below 3.463		At 10
LL2		Read 1.681				
D			To .681	Read 2.358 = atmos.	To 2.358	Read 34.66 lb. per sq. in. = mean effective pressure.
CI					Set 14.7	

SETTINGS INVOLVING THE LOG-LOG SCALES   193

(Continued)	CF	Above 144	Set 550	Above 1
	DF	Read 4990 lbs. per square foot × 1 foot stroke = 4990 ft. lbs.	To 4990	Read 9.08 HP

71. Airways:

To find the diameter of a round airway to pass the same amount of air as a square airway, the length and power remaining the same.

$$D^3 = \sqrt[5]{\frac{A^3 \times 3.1416}{.7854^3 \times O}},$$

where D = the diameter of the round airway.
A = the area of the square airway.
O = the perimeter of square airway.

For slide rule purposes this may be written:

$$D = \sqrt[15]{\frac{A^3 \times 3.1416}{(.7854)^3 \times O}}.$$

D	Under A	Under .7854		
K	Read A^3	Read $(.7854)^3$		Read $\frac{A^3 \times 3.1416}{(.7854)^3 \times O} = x$
DF			To A^3	
CF			Set $(.7854)^3$	
CI				Above O
(Continued)		LL	To x	Find D
		C	Set 15	Below 1

72. Flow of Steam in Pipes:

$$d = 0.5374 \sqrt[5]{\frac{Q^2 l}{h}}$$

where d = internal diameter of pipe in inches.
Q = quantity of steam in cubic feet per minute.
l = length of pipe in feet.
h = height of a column of steam, of the pressure of steam at entrance, which would produce a pressure equal to the difference of pressures at the two ends of the pipe.

194 TYPICAL PROBLEMS AND SLIDE RULE SETTINGS

D	Ind. to Q				To $\sqrt[5]{\dfrac{Q^2 l}{h}}$	Read Ans.
B	h to Ind.	At l				
A		Read $\dfrac{Q^2 l}{h}$				
C			Set 5	At index	Set index	At .5374
LL			To $\dfrac{Q^2 l}{h}$	Read $\sqrt[5]{\dfrac{Q^2 l}{h}}$		

73. Effects of Bends and Curves in Pipes:

Loss of head in feet $= \left\{ .131 + 1.847 \left(\dfrac{r}{R}\right)^{\frac{7}{2}} \right\} \times \dfrac{v^2}{64.4} \times \dfrac{a}{180}$

r = internal radius of pipe in feet.
R = radius of curvature of axis of pipe in feet.
v = velocity in feet per second.
a = central angle, or angle subtended by bend.

C	Set R	Below index				
D	To r	Read $\dfrac{r}{R}$				
LL0			To $\dfrac{r}{R}$	Read $\left(\dfrac{r}{R}\right)^7$	To $\left(\dfrac{r}{R}\right)^7$	Read $\left(\dfrac{r}{R}\right)^{\frac{7}{2}}$
B			Set index	Above 7	Set 2	Above index

(Continued)

	C	Set index	Below 1.847	
	D	To $\left(\dfrac{r}{R}\right)^{\frac{7}{2}}$	Read $1.847 \left(\dfrac{r}{R}\right)^{\frac{7}{2}} = x.$	

(Continued)

D	Indicator to v			
B	Set 64.4 to Ind.	Ind. to $.131 + x$	180 to Ind.	Ind. to a
A				At Ind. read Ans.

SETTINGS INVOLVING THE LOG-LOG SCALES   195

74. **Loss of Pressure and Head in Rubber-Lined Smooth 2½ Inch Hose:**

$$P = \frac{lg^2}{4150\, d^5} \qquad h = \frac{lg^2}{1801\, d^5}$$

P = pressure lost by friction in lbs. per sq. in.
l = length of hose in feet.
g = gallons of water discharged per minute.
d = diameter of hose in inches.
h = friction-head in feet.

C	Set index	Below 5		Indicator to g		
LL	To d	Read d^5				
A			To 1			Read Ans.
B			Set 4150 or 1801		d^5 to Indicator	At Index

75. **Relation of Diameter of Pipe To Quantity of Water Discharged:**

$$d = .239 \left\{ \frac{Q}{\left(\frac{h}{L}\right)^{\frac{1}{2}}} \right\}^{.387}$$

d = diameter of pipe in feet.
Q = quantity of water discharged in cubic feet per second.
h = head in feet.
L = length of pipe in feet.

A	To L					
B	Set h					
C		Below Q	Set Index	Under .387	Set index	Under .239
D		Find $\frac{Q}{\left(\frac{h}{L}\right)^{\frac{1}{2}}} = x$			To $x^{.387}$	Read Ans.
LL			To x	Find $x^{.387}$		

76. **Density and Volume of Saturated Steam:**

$$V = \frac{330.36}{p^{.941}} \qquad D = \frac{p^{.941}}{330.36}$$

V = volume.
p = pressure in lbs. per square inch.
D = density.

196 TYPICAL PROBLEMS AND SLIDE RULE SETTINGS

C	Set Index	Below 941	Set $p^{.941}$	Below index	Read D
LL	To p	Read $p^{.941}$			
D			To 330.36	Read V	Opposite Index

77. Total Heat of Superheated Steam:

$$H = 0.4805(T - 10.38\, p^{\frac{1}{4}}) + 857.2.$$

H = total heat of superheated steam.
T = temperature in degrees F. $+ 460.7$.
p = pressure in lbs. per square foot.

C	Set 4	Below 1	Set 1	Below 10.38	Set 1
LL	To p	Read $p^{\frac{1}{4}}$			
D			To $p^{\frac{1}{4}}$	Read 10.38 $p^{\frac{1}{4}}$	To $T - 10.38\, p^{\frac{1}{4}}$

(Continued)

C	Under 0.4805	To answer
LL		add
D	Read $.4805(T - 10.38\, p^{\frac{1}{4}})$	857.2

78. Loss of Pressure of Steam Due to Friction:

Loss of power, expressed in heat units, due to friction:

$$H_f = \frac{W^3 f l}{10\, p^2 d^5}$$

W = weight in lbs. of steam delivered per hour.
f = the coefficient of friction of the pipe.
l = length of pipe in feet.
p = absolute terminal pressure.
d = diameter of pipe in inches.
   f is taken as from .0165 to .0175.

D	Under W				
K	Read W^3				
C		Set Index	At 5	Set p	
LL		To d	Read d^5		
A				To W^3	
B					Indicator to f

SETTINGS INVOLVING THE LOG-LOG SCALES

(Continued)	A		At Indicator read Answer.
	B	Set d^5 to Indicator	Indicator to 1

79. Mean Pressure of Expanded Steam:

$$P_m = p_1 \frac{1 + \log_e R}{R}$$

P_m = absolute mean pressure.
p_1 = the absolute initial pressure taken as uniform up to the point of cut-off.
R = absolute initial pressure.

C	Set Index	Read $\log_e R$	Set R	Indicator to p_1
LL	To e	Above R		
D			To $1 + \log_e R$	At Indicator read Ans.

80. Relative Efficiency of One Lb. of Steam With and Without Clearance:

Back pressure and compression not considered.

Mean total pressure $= p = \dfrac{P(l + c) + P(l + c)\log_e R - Pc}{L}$

P = initial absolute pressure in lbs. per sq. in.
R = actual ratio of expansion $= \dfrac{L + c}{l + c}$
l = period of admission measured from beginning of stroke.
c = clearance in inches.
L = length of stroke in inches.

C	Set $(l + c)$	At index	Set index	Indicator to $(l+c)$	Index to Ind.
D	To $(L + c)$	Read R			
LL			To R		

(Continued) C	Indicator to P	Index to $(l + c)$	At P
D	Read $P(l + c)\log_e R$		Read $P(l+c)$

(Continued)

C	Set L	At Index
D	To $P(l + c) + P(l + c)\log_e R - Pc$	Read p

198 TYPICAL PROBLEMS AND SLIDE RULE SETTINGS

81. **Diameter of Piston Rods:**
$$d = \sqrt[4]{\frac{D^2pL^2}{a}} + \frac{D}{80}$$

D = diameter of cylinder in inches.
p = maximum unbalanced pressure in lbs. per sq. in.
L = length in feet.
a = 10,000 and upward, increasing with decrease in speed of engine.

C				At L	Set 4	At Index
A				Read $\frac{D^2 p L^2}{a}$		
B	Set a	Ind. to p	Index to Indicat.			
D	To D					
LL					To $\frac{D^2 p L^2}{a}$	Read $\sqrt[4]{\frac{D^2 p L^2}{a}}$

(Continued)

C	Set 80	At index
A		
B		
D	To D	Read $\frac{D}{80}$
LL		

82. **Journal Friction:**
Coefficient when shaft is revolving
$$= (0.2 \text{ to } 0.3) \frac{\sqrt[5]{\text{vel. in ft. per min.}}}{\sqrt[5]{\text{press. in lbs. per sq. in.}}}$$

NOTE: This coefficient is for ordinary temperatures, pressures and speeds, with journals and bearing in good condition and well lubricated.

C	Set 5	Opposite index
LL	To vel. in ft. per min.	Find $\sqrt[5]{\text{vel. in ft. per min.}}$

(Continued)

C		At 0.2 or 0.3
D	To $\sqrt[5]{\text{vel. in ft. per min.}}$	Read Answer
A	Set press. in lbs. per sq. in.	

SETTINGS INVOLVING THE LOG-LOG SCALES 199

83. **Storm Flow-Off:**

Cubic feet per second = A coefficient according to judgment E × Average cu. ft. of rainfall per sec. per acre during heaviest fall N × $\sqrt[4]{\dfrac{\text{Average slope of ground in feet per 1,000 feet} = S}{\text{No. of acres drained} = T}}$

C	Set T	Opposite 1	Set 4	Opposite 1		At N
D	To S	Read $\dfrac{S}{T}$			To E	
LL			To $\dfrac{S}{T}$	Read $\sqrt[4]{\dfrac{S}{T}}$		
CI					Set $\sqrt[4]{\dfrac{S}{T}}$	Read Answer

84. **Turbine Discs:**

$$Y = Y_a e^{-\dfrac{uw^2x^2}{2t}} \quad \text{or} \quad Y = \dfrac{Y_a}{e^{\dfrac{uw^2x^2}{2t}}} \text{ (according to Stodola).}$$

Y = thickness at radial distance x
Y_a = thickness of disc carried to shaft center.
x = radial distance of a point from the axis.
u = specific mass.
w = angular velocity.
t = radial and tangential stress per unit of area.

C	Set index	Ind. to w	1 to Ind.	Ind. to x	
A	To u				
B					

(Continued)

C				Read Ans.
A				
B	2 to Index	Ind. to Index	t to Ind.	At Index

Any number of values are thus found for $\dfrac{uw^2x^2}{2t}$, by giving x a

200 TYPICAL PROBLEMS AND SLIDE RULE SETTINGS

number of different values. With the indexes of C and LL coinciding, read under each value of $\dfrac{uw^2x^2}{2\,t}$ on C the corresponding value of $\dfrac{uw^2x^2}{e\;2\,t}$ on LL.

From that point onward the solution is one of simple division on the C and D scales.

85. **Catenary Curve:**

$$y = \frac{a}{2}\left(e^{\frac{x}{a}} + e^{-\frac{x}{a}}\right)$$

C	Set a	At Index	Set 1	At $\dfrac{x}{a}$		
D	To x	Read $\dfrac{x}{a}$			To e	
LL3			To e	Read $e^{\frac{x}{a}}$	Set 1	Read $e^{-\frac{x}{a}}$
LL0						At $\dfrac{x}{a}$

Since $e^{-\frac{x}{a}}$ is the reciprocal of $e^{\frac{x}{a}}$, the corresponding value of $e^{-\frac{x}{a}}$ may be found on CI directly over the value of $e^{\frac{x}{a}}$ on C.

The remainder of the computation is a simple addition; coupled with a simple division, using the CD scales.

Example: Let $a = 8$, and $x = 0, 1, 2, 3, 4, 5, 6, 7, 8$.

x	$\dfrac{x}{a}$	$e^{\frac{x}{a}}$	$e^{-\frac{x}{a}}$	$e^{\frac{x}{a}} + e^{-\frac{x}{a}}$	$\dfrac{a}{2}\left(e^{\frac{x}{a}} + e^{-\frac{x}{a}}\right) = y$
0	0.	1.	1.	2.	8.
1	.125	1.133	.882	2.015	8.06
2	.25	1.284	.779	2.063	8.252
3	.375	1.455	.687	2.142	8.568
4	.5	1.649	.606	2.255	9.02
5	.625	1.869	.535	2.404	9.616
6	.75	2.12	.472	2.592	10.368
7	.875	2.4	.416	2.816	11.264
8	1.	2.718	.368	3.086	12.344

SETTINGS INVOLVING THE LOG-LOG SCALES 201

86. **Transmission Lines:**

The capacity in Electro-static units of long transmission lines is given by:
$$C = \frac{l}{2 \operatorname{Log}_e \frac{l_1}{r}}$$

C	Set r	At index	Set l		Read answer
D	To l_1	Read $\frac{l_1}{r}$			At l
LL			To e	Ind. to $\frac{l_1}{r}$	
CI				Set 2 to Ind.	

If length l is required the formula becomes
$$l = 2\, C \operatorname{Log}_e \frac{l_1}{r}.$$

C	Set r	At index		At 2
D	To l_1	Read $\frac{l_1}{r}$		Read Answer.
LL			To $\frac{l_1}{r}$	
CI			Set C	

87. **Hysteresis Loss:** $\quad W_h = nB^{1.6}$

C	Set 1	At 1.6	Set 1	At n	
LL	To B	Read $B^{1.6}$			
D			To $B^{1.6}$	Read $W_h =$ Answer.	

88. **The Resistance of Dielectrics:**
$$R = \frac{It}{C(\operatorname{Log}_e E_1 - \operatorname{Log}_e E_2)}$$

LL	At E_1	At E_2		
D	Read $\text{Log}_e E_1$	Read $\text{Log}_e E_2$	To It	Read R
C			Set C	
CI				At($\text{Log}_e E_1 - \text{Log}_e E_2$)

89. Photometry and Light Transmission:

The general formula is as follows:

$$T' = T^{\frac{l'}{l}}$$

in which

T = transmission of the reference medium
l = length of path or thickness
T' = transmission sought for length l'.

(i) *Example:* A particular glass 2.4 cm. thick transmits, exclusive of the reflection loss, 88 per cent of a certain wave length. What will be the transmission of 6.4 cm. of the same glass?

$$T' = .88^{\frac{6.4}{2.4}}$$

LL0	To .88	Find .711. Answer
B	Set 2.4	Over 6.4

(ii) *Example:* A glass ray filter 3.2 mm. thick transmits, exclusive of the reflection loss, 52 per cent of violet light. What must be the thickness to transmit 75 per cent of the same light?

$$.75 = .52^{\frac{l'}{3.2}}$$

LL0	To .52	Opposite .75
B	Set 3.2	Find 1.4 mm. Answer

The light transmission measured by the photometer is the total, which includes the reflection loss at the two surfaces.

SETTINGS INVOLVING THE LOG-LOG SCALES 203

From photometric readings of total the relative transmission only can be computed by the formula.

$$T' = \left(\frac{T_1(n+1)^4}{16\, n^2}\right)^{\frac{l'}{l}}$$

In which
n = Refractive index of medium.
T' = Transmission sought for length l'.
T_1 = Total transmission of length l.

(iii) *Example:* A glass plate 1.5 cm. thick, refractive index 1.52, has a total transmission of 72 per cent of visible light. What will be the transmission only of the same glass, but 2.5 cm. thick?

$$T' = \left(\frac{.72 \times 2.52^4}{16 \times 1.52^2}\right)^{\frac{2.5}{1.5}}$$

A	To 7.2			Read .785		
B	Set 16				Set 1.5	At 2.5
C		Ind. to 2.52	1.52 to Ind.	At 2.52		
LL0					To .785	Find .668 Answer

CHAPTER V

Special Forms of the Slide Rule

39. Introduction. — The slide rule described and explained in Chapter II is the basic standard form from which most other modern forms are derived. The rules treated in Chapter III are in general use in almost all the world, the names given in this book being those used in the United States. Beside these, which have become more or less standardized, there are many other forms of the slide rule which are specially constructed for certain purposes or have special scales for particular types of problems or calculations. A few of these will be described briefly in this chapter.

The various special forms of rules may of course take any desired construction form and they are actually of the greatest variety. In general, however, the aim seems to be to adhere to some more or less standard construction and to adapt the rule to its particular purpose by variations of the usual scale arrangements and the addition of new scales for the special calculations for which the rule is designed. In this way most of these special rules take the mechanical form of the Mannheim and duplex

INTRODUCTION 205

rules, or they are made circular with scales on the circumferences or cylindrical with scales placed helically or parallel to the axis.

In what follows the rules discussed will be classified according to mechanical construction and two or three of each form, having different scales or scale arrangements, will be described.

Most of the rules described in this chapter carry two or more of the standard Mannheim scales which have already been described in detail. In the case of the other scales the special types of problems to which they are adapted limits their use to those who are particularly interested in such problems, or the special forms of their construction and scale division requires that a description, in order to be of value to the user, be given in considerable detail. In either case the general student and user will not be particularly interested in such descriptions, and the specially interested user will prefer to refer to the instructions issued with each form of rule and covering its operation in detail.

For these reasons the treatment given here will consist of a brief description of the construction and scales, accompanied in each case by a cut picturing the rule, and a brief statement of its uses, without any pretense to the fullness of the treatments given for the standard rules in Chapters II and III.

206   SPECIAL FORMS OF THE SLIDE RULE

40. Slide Rules of Mannheim Form with Special Scales. — Two rules of this type will be described briefly; they are the Roylance Electrical Slide Rule and the Keuffel & Esser Stadia Slide Rule. The Roylance Electrical Rule is illustrated in Fig. 48 below which shows

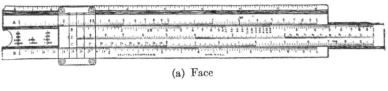

(a) Face

(b) Edge

FIG. 48.   ROYLANCE ELECTRICAL SLIDE RULE

the 8-inch rule. Mechanically this rule is exactly like the regular Mannheim rule and it has the same scales as the Mannheim rule and in addition the inverted (CI) scale of the polyphase rule.

On the regular C scale is a gauge point at 746 which is used for horse-power and kilowatt conversions. The regular B scale is complete, but the graduations beginning at 9.6 and ending at 20 serve also as a temperature scale in the determination of electrical resistance of wires at different temperatures. In this range the point 9.6 is specially marked, the middle index is marked in red, 11 is marked 25, 15 is marked 135, and 30 is marked 260. Consequently this portion of the B scale serves a double purpose. On A2 near 3 is a guage point marked "N" which is used in determining the weight per thousand feet of bare copper wire. In addition to the usual indicator hair line the runner carries two other such lines equidistant from it on either side. The distance between these outer lines is equal to the distance from the gauge point $\frac{1}{4}\pi = .7854$ to the right index on A2. Therefore if the runner is set with the right hand line at a diameter on D the circular area is at the left line on A. This device is useful in connection with wire size calculations.

The additional scales consist of a double scale on the lower edge of the stock, shown in Fig. 48 (b), and a multiple scale of gauge points on the stock under the slide, as shown in Fig. 48 (a). The scale on the edge gives the Brown & Sharpe gauge (American Wire Gage) numbers of wires when the central hair line of the runner is set on the diameter at D. The scales in the groove under the slide show the ampere carrying capacities of insulated wires and cables. The upper row of figures applies to rubber covered wire, the second row to weatherproof wire, the third to rubber covered cable and the fourth to weatherproof cable. The numbers on these scales are in line with the wire gauge scale divisions on the edge and are read in connection with those by means of the indicator. For wires the gauge numbers are regular B. & S. For cables they are in terms of 100,000 circular mils; thus gauge No. 8 reads 800,000 c.m.; No. 14 reads 1,400,000 c.m., etc.

In addition to all the calculations possible with the Mannheim and many of those possible with the polyphase rules, therefore, the scales dealing with special electrical wiring problems make it possible to read directly at one setting of the indicator for any size of copper wire, the diameter in mils and inches, the section area in c.m. and sq. in., weight in pounds per thousand feet of bare wire, and resistance in ohms per thousand feet at any temperature in degrees Centigrade. In addition to these calculations and the power conversions, any of these data may of course be used in any calculation possible on the Mannheim rule or involving any of the standard electrical formulas.

The Keuffel & Esser Stadia slide rule, which is also of the same mechanical construction as the Mannheim rule,

FIG. 49. KEUFFEL & ESSER STADIA SLIDE RULE

is shown in Fig. 49. On the stock are two scales (see figure) both graduated the same as the standard A scale. The lower scale is marked to read 1 to 10 to 100 and indicated as "A." The upper one, indicated as "R," is marked to read 10 to 100 to 1000. On the back of the slide is a "B" scale like the "A" on the stock and when the slide is reversed these two are used together for the ordinary purposes of the C and D scales. The rule is made in 16 and 20 inch lengths and each half of "A" and "B" is thus equivalent to the 8 or 10 inch C or D scale.

On the front of the slide, shown in the figure, are three scales called H, V and HC. The upper scale graduations are the same as those of the ordinary S scale and the markings are the same up to 40°. The remainder of this scale, instead of reading 50, 60, 70, 90°, reads from the right index toward the left as 0, 20, 30, 40, 45°. The left-hand portion is called "V" and the right-hand portion "H." The lower scale on the slide reads from left to right throughout its full length and is marked 4' to 5° 43', which is the part of the angle scale (less than 45°) which does not appear on the ordinary T scale. This scale is also called "V." The third scale on the slide, between the upper and lower, is called "HC" and reads left to right from 1° 49' to 18° 15'. These three scales of angles are used in connection with R and A on the stock to read the horizontal and vertical distances in stadia surveying when the vertical angle and stadia rod readings are known, by one setting of the slide.

This rule is graduated so that for the values of vertical

210   SPECIAL FORMS OF THE SLIDE RULE

angle commonly met in stadia surveying the 20-inch rule gives the distances correct to the nearest 0.05 foot. The beveled edge of the projecting "lip" of the stock (shown in Fig. 11 (a)) is graduated in inches and tenths to serve as a plotting scale for distances.

41. Duplex Slide Rules with Special Scales. — The duplex slide rule described in Article 27 lends itself particularly to special scale arrangements due to the large

(a) Front

(b) Back

Fig. 50. Surveyor's Duplex Slide Rule

space available and to the fact that both faces are immediately accessible without the necessity of removing and reversing the slide. On this account duplex rules with many special forms of scale have been devised. Three such rules will be considered here.

The first of these is the Surveyor's Duplex Slide Rule, shown in Fig. 50. The back of this rule (Fig. 50 (b)) carries the regular A, B, C, D and CI scales and also the

H and V scales of the stadia slide rule described in the preceding Article.

The Front face (Fig. 50 (a)) carries the regular S and T scales (here marked "Sin" and "Tan") and also a cosine scale graduated in degrees. These are used in computing course latitudes and departures. On this face are five other scales arranged for the determination of the meridian from the transit readings of a direct solar observation. Since the operations involved in the use of these scales are somewhat complicated they will not be described here and the specially interested reader is referred to the book of instructions supplied with the rule. The rule is 20 inches long and when used in connection with the transit instrument with vertical circle and solar attachment, permits the making in the field of all the computations of surveying.

The second special form of duplex rule to be considered here is the Chemist's Duplex Slide Rule designed by Dr. R. Harmon Ashley. It is illustrated in Fig. 51.

On the front of the rule are the regular C and D scales in their usual positions and on the back are the CI and D scales. Beside the ordinary operations with these scales the direct and inverse proportions common in chemical calculations find them of particular convenience.

On the front of the rule are duplicate special scales or sets of gauge points giving the chemical formulas of the

212 SPECIAL FORMS OF THE SLIDE RULE

more common acids, bases and salts; one is on the stock and one on the slide. On the back are similar and similarly located scales giving the symbols of the chemical elements and the formulas of the more common oxides. The gauge points on these four scales are arranged in such order and sequence that by using the scales directly in connection with one another the reaction products

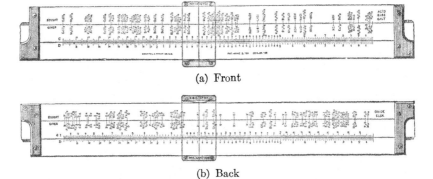

(a) Front

(b) Back

Fig. 51. Ashley Chemist's Slide Rule

"sought" (see figure) may be found when the reagents are "given." By using these scales in connection with C, D and CI the computations of gravimetric and volumetric analysis, stoichiometry, equivalents, percentage composition, volume of a gas from a given weight of substance at different temperatures and pressures, and other analogous problems, may be carried out by means of very simple operations.

DUPLEX SLIDE RULES WITH SPECIAL SCALES 213

A third type of duplex rule is the Power Computing Slide Rule (Fig. 52) designed especially for use in connection with the design, testing and operation of steam, gas and oil engines.

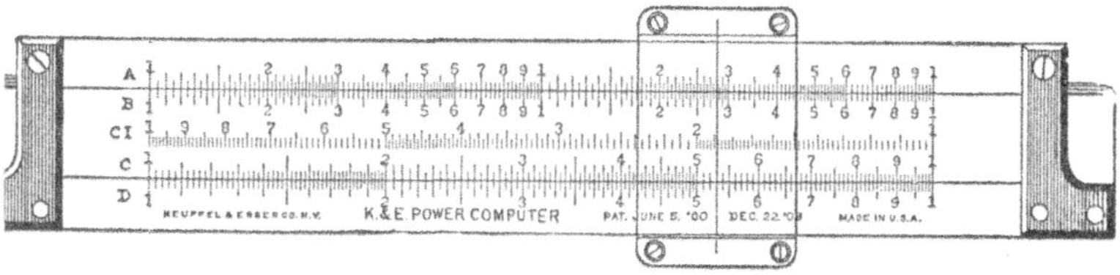

(a) Front

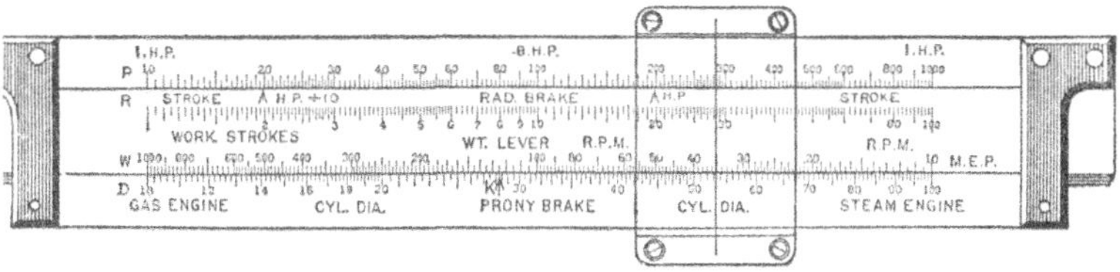

(b) Back

FIG. 52. POWER COMPUTING DUPLEX SLIDE RULE

The front face carries the regular A, B, C, D and CI scales and can therefore be used for all the ordinary forms of calculations.

On the back are the special scales P, R and W and a regular D scale which is marked 10 to 100. On all these scales are marked special gauge points and each scale is also marked to indicate the type of computation for which it is to be used. The computations for which these scales

214   SPECIAL FORMS OF THE SLIDE RULE

and gauge points are specially adapted are those involving the power, steam or fuel consumption, dimensions and Prony brake tests of engines, and they also give certain standard data and conversion factors. Some of these computations are for speed, stroke, cylinder dimensions, brake and indicated horse-power, mean effective pressure, etc.

In ordinary calculations the position of the decimal point is usually known or obvious on inspection. In gas and steam engine work the scales as marked on this rule take care of the decimal point when the factors and results are within the numbered range. Any other values may be handled by shifting the decimal point as required. In Prony brake work, if all the factors taken are as numbered on the scales, the brake horse-power read at the H.P. index must be divided by 100.

42. **Circular Slide Rules.** — As seen in Chapter I the circular slide rule, having the scales laid off on concentric circular discs or rings of which one turns upon or within the other, was invented in England by William Oughtred before 1630 and independently reinvented in France by Jean Clairaut in 1716. The circular rule, while never very popular in England where most of the early development of the slide rule took place nor in the United States where most of the recent developments have

occurred, has since 1716 been much used in France and a widely used form of this type is the "Charpentier Calculator." The use of this rule is beginning to spread in the United States.

The Charpentier Slide Rule is shown in Fig. 53. It is of the general form of a watch, about $2\frac{1}{2}$ inches in diameter, and is made entirely of metal. The opposite faces are plane and it is in effect a single disc. The "stock" of the rule is a ring and the "slide" is a smaller disc which turns inside this disc with faces flush, being rotated on the center and set by means of a radial handle which carries at the end a "watch chain ring." On the front of the instrument (shown in the figure) is the C scale on the rotary slide and the D scale on the outer ring. The A and B scales in the usual double form are not included but an inner scale in two parts on the slide takes the place of the usual C scale ($\sqrt{1} = 1$ to $\sqrt{10} = 3.162$ to $\sqrt{100} = 10$) when the C scale is here used as the ordinary B scale. Thus the C scale and the two inner scales of this instrument are used together in the same way as the ordinary B and C scales.

On the back of the instrument are the usual L scale on the outer ring and the S and T scales on the disc slide. These are used in connection with the corresponding scales on the front by means of the pointers shown in the figure. The C and D scales are 6 inches in length, thus being

216   SPECIAL FORMS OF THE SLIDE RULE

somewhat better than those of the ordinary 5-inch straight slide rule.

Fig. 54 illustrates another form of circular slide rule which is of American origin and known as the "Sperry Pocket Calculator." The instrument has the form of a watch with an engraved glass-covered metal dial on each

FIG. 53. CHARPENTIER CIRCULAR   FIG. 54. SPERRY POCKET
           SLIDE RULE   CALCULATOR

side. Each dial has an index hand and a stationary pointer which together correspond to the runner (indicator) of the ordinary form of rule and which are set by the "stem" as in a watch. A second stem nut revolves the "slide" dial. The scales run twice around in the form of a spiral and the main logarithmic scales (C, D) are thus about $12\frac{1}{2}$ inches long, being therefore somewhat better than the usual 10-inch rule. The instrument has all the Mannheim scales except S and T.

The advantages of these rules consist in their compactness and the fact that the scales are endless and there is no running off the scale and interchanging indexes as is the case with a straight rule.

43. Cylindrical Slide Rules. — The two cylindrical slide rules most used in the United States are the famous Thacher Rule (see Article 7) and Fuller's Slide Rule. These are illustrated in Figs. 55 and 56 respectively.

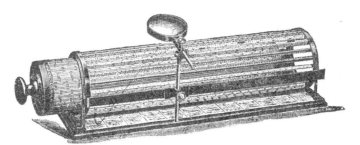

Fig. 55. Thacher Cylindrical Slide Rule

As shown in the illustration, Fig. 55, the Thacher Slide Rule consists of a cylinder which is 4 inches in diameter and 18 inches long and which revolves in an open framework composed of 20 bars running lengthwise and held radially edgewise between two metal rings. The cylinder bears scales corresponding to the logarithmic scales of the

218   SPECIAL FORMS OF THE SLIDE RULE

ordinary slide rule, which are duplicated on the exposed sides of the parallel bars. The cylinder slides and rotates inside the framework and any part of any one of its scales can be set at any part of any of the scales on the bars. The cylinder is provided with a knob at each end for sliding and rotating it, and the framework bears a 3-inch reading glass mounted on an adjustable stand which slides along the length of the instrument on a brass bar. The glass may thus be set at any part of the instrument and is also adjustable for focus.

The instrument has all the scales of the polyphase Mannheim slide rule and the equivalent total length is such that it may be read to four and five figures without the glass while with the glass it may be read to five and six figures. The scale arrangement is such that any of the usual slide rule operations involving as many as three given quantities are performed by one setting, while any number of values of an algebraic function composed of two constants and one variable may be found with a single setting.

The Thacher rule is by reason of its construction and range particularly suited to desk and drafting table use in engineering research and design offices and laboratories.

As shown in Fig. 56 Fuller's Slide Rule consists of a hollow outer cylinder (a) which slides and rotates on an inner cylinder (b), the inner cylinder being provided with a handle (c). A single logarithmic scale nearly 42 feet long is wound helically on the outer cylinder and there are three indexes, one of which is on the fixed pointer (d) attached to the handle. A brass tube (e) slides inside the cylinder (b) and attached to it is a brass strip or bar (f) parallel to the axis. This bar carries two indexes

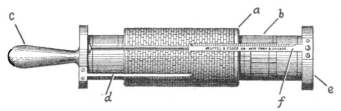

FIG. 56. FULLER CYLINDRICAL SLIDE RULE

whose distance apart is the axial length of the helical scale on (a), and a scale of equal parts for the finding of logarithms. On the cylinder (b) are printed a number of tables, gauge points and settings.

In using the ordinary slide rule any operation involving multiplication, division, powers and roots (special forms of multiplication and division), and proportion can all be looked upon as the solutions of proportions (see Articles 17 and 18). These proportions are set up on Fuller's Slide Rule by setting one number on the scale to the fixed index and the movable indicator arm to the second member on the scale, and then bringing the third member on the scale to the fixed index and reading the fourth on the scale at the movable indicator. Hence the Fuller rule requires re-setting each time the third member of a proportion changes and it therefore does not give a series of equal ratios at sight in any one setting as do the C, D or A, B scales of the straight rule. Due to the extraordinary length of the scale however, it can be read with very great precision.

A cylindrical slide rule which has recently appeared takes the simple form of the popular and convenient refillable pencil. It carries only the C and D scales and these are arranged parallel to the axis of the pencil, one on the body and the other on a sliding sleeve. The pencil is about 6 inches long so that the rule is about equivalent to the ordinary 5-inch slide rule.